LEARNING MATHE
THROUGH THE TI

JOHN S. BERRY, EDWARD GRAHAM and
ANTHONY J. P. WATKINS
Centre for Teaching Mathematics
University of Plymouth

Chartwell-Bratt   Studentlitteratur

Other books in the "Learning through Computer Algebra" Series
Published by Chartwell-Bratt. Series editor: John Berry.

"Learning Mathematics through DERIVE", J. S. Berry, E. Graham, A. J. P. Watkins
"Learning Numerical Analysis through DERIVE", T. Etchells, J. Berry
"Learning Linear Algebra through DERIVE", B. Denton
"Learning Differential Equations through DERIVE" by B. Lowe, J. Berry
"Learning Modelling through DERIVE", S. Townend, D. Pountney.

British Library Cataloguing in Publication Data
A catalogue record for this book is available from the British Library

http://www.studli.se/chartwell.html

Chartwell-Bratt (Publishing and Training) Ltd
ISBN 0-86238-489-3

Printed in Sweden
Studentlitteratur, Lund
Art.no. 6284
ISBN 91-44-00482-6

Printing/Year   1 2 3 4 5 6 7 8 9 10 | 02 01 2000 99 98 97

Contents

Contents

Contents

Series Preface

This series is designed to encourage the teaching and learning of various courses in mathematics using symbolic algebra. Each text in the series will include the 'standard theory' appropriate to the subject, worked examples, symbolic algebra exercises and Activities so that students have a sound conceptual base on which to build. It is through these Activities that the learning of mathematics through investigations and discussion is encouraged. The aims of each Activity are to provide an investigation to introduce a new topic and to introduce the symbolic algebra commands for using as a manipulator in mathematical problem solving. Mathematics teaching and learning is changing as new technology becomes more readily available and these texts are written to support the new methods.

John Berry
Centre for Teaching Mathematics
University of Plymouth

Preface

The acceptance of calculator technology as an integral part of the mathematics curriculum in schools, colleges and universities is a continuing subject for discussion with the level and heat of the debate rising and falling periodically. It's always the same issues and questions that are asked but the level of the curriculum is different. "When pupils use calculators they will not be able to do 'such and such'. Should we allow calculators in the classroom before pupils can do maths by hand? Should we allow calculators in the examinations?"

Twenty-five years ago the basic number calculator arrived in the primary school and the debate about children's ability at mental arithmetic schools started (and continues today!). Ten years later the scientific and then graphics calculators arrived in secondary schools. Initially the latter were banned from public examinations. Now they are accepted as an integral part of the teaching and examining at GCE Advanced level.

Now the arrival of the TEXAS TI-92 has opened up the debate once more. The TI-92 guidebook states that the TI-92 has 'the math power of a computer with the independence of a graphing calculator'. With such powerful tools as the TI-92 we could see more of a *revolution* in changes to the curriculum and learning styles instead of the *evolution* of the past twenty years.

This text demonstrates how symbolic algebra software, such as that in the TI-92, can be used to introduce new concepts to students providing conceptual understanding and skill development. This text is based on the result of several years experience of using symbolic algebra software in our teaching.

The content of this text consists of the basic mathematics taught in nearly all first courses in mathematics for science and engineering. It is based on the experience of the three authors of teaching such material for many years. The text falls naturally into three sections: functions, calculus and algebra. Chapter 1 is essentially a review showing how to use the TI-92 to explore mathematical concepts and to use it in modelling and mathematical problem solving. In chapters 2 and 3 we introduce the exponential, logarithmic and trigonometric functions. Chapter 4 is about sequences and series. Numerical methods are integrated into the text at the appropriate points through chapters 5 and 8. The bulk of the calculus takes up a third of the text in chapters 6 and 7. For the engineer and scientist these chapters provide the mathematical tools for modelling many real situations. Chapters 9 and 10 look briefly at an introduction to matrices and complex numbers. In schools and colleges in the United Kingdom, this material is taught as part of GCE Advanced level and corresponding BTEC and GNVQ courses and in many Universities the material forms part of a foundation course in science and engineering.

The text contains the 'standard theory', worked examples, exercises and TI-92 Activities. It is through these TI-92 Activities that the learning of mathematics through investigation and discussion is encouraged. The aims of each TI-92 Activity are to provide an investigation to introduce a new topic (often before a more formal approach) and to introduce the commands for using the TI-92 as a manipulator in mathematical problem solving. The TI-92 Activities could be used as a resource for workshop sessions with large groups, for self-learning a new topic, or reviewing a topic taught in the more traditional lecture/lesson.

We would encourage students to work cooperatively in small groups (e.g. pairs) through the TI-92 Activities, discussing the outcomes of each part. The teacher should interact with the student groups checking their conjectures and providing further tasks, as necessary, to reinforce the student conceptual understanding and skill development.

What makes this text different from the many others on the market at this level is the use of the TI-92 calculator. This powerful tool does for algebraic manipulation what the scientific calculator does for arithmetic. However a word of warning! The authors believe that the introduction of symbolic algebra into the learning classroom *does not replace* the need to learn and understand algebraic manipulation. Our experience suggests that the *best* users of symbolic algebra are those students who are good at algebraic manipulation and understand the concepts behind algebra and calculus. The text attempts to blend the traditional skills with the power of the TI-92. To encourage good practice many of the exercises should be done 'by hand' first and then the results checked using the TI-92.

This book has been adapted from 'Learning Mathematics through DERIVE'. Many of the TI-92 Activities are new Investigations which exploit the TI-92's facilities to introduce new concepts to students. It is a pleasure to acknowledge the work of Jonathon Martin, one of our undergraduate students, who worked through the DERIVE book producing a first draft for this TI-92 book.

We are extremely grateful to Sharon Ward for her skill and meticulous care in producing the camera ready copy for this text; to Jenny Sharp for producing the artwork which has helped enhance the finished quality of the text and to John Monaghan for helpful comments on the TI-92 Activities.

We would welcome comments from teachers and students for improving this text. Since this is one of the first textbooks to fully integrate the TI-92 calculator into the learning of mathematics it is inevitable that the mix of the TI-92 Activities and standard theory could be improved. If you have constructive criticisms to help the writing of a second edition please write to the authors in Plymouth.

We hope that the teacher and learner will enjoy their mathematical activities using the TI-92.

John Berry
Ted Graham
Tony Watkins

Centre for Teaching Mathematics
The University of Plymouth
Drake Circus, Plymouth
Devon PL4 8AA, UK
Telephone/FAX 01752 232772

Hints for the TI-92 Activities

We have assumed that the reader is familiar with using the TI-92. Before beginning each TI-92 Activity you are recommended to clear the HOME screen, the Y = Editor and any use-defined variables by pressing the following keys

> **F6 ENTER**
> **F1** select 8:clear Home **ENTER**
> **CLEAR**
> ◆ **W F1** SELECT 8:clear Functions **ENTER**

In each TI-92 task we have used bold text to identify the keys to be pressed. For example, **MODE** means 'press the key labelled MODE'.

Unless stated otherwise at the start of each TI-92 Activity set the MODE of your TI-92 to RAD and AUTO.

To evaluate an exact expression as a decimal press ◆ **ENTER**. The following HOME screen dump shows the effect of using **ENTER** and ◆ **ENTER** when in AUTO mode.

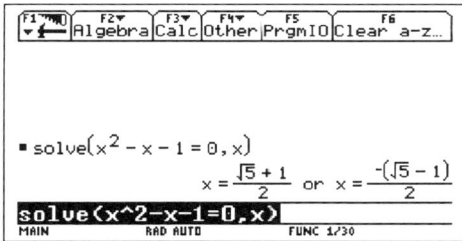

an exact answer   an approximate answer

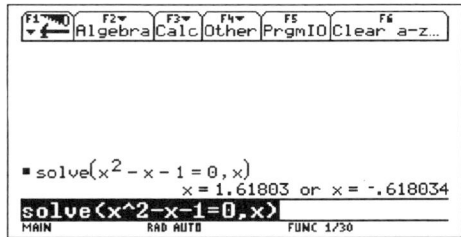

1

Introducing functions

1.1   LINEAR LAWS

TI-92 ACTIVITY 1A

(A)   The aim of this activity is to familiarise you with the Data/Matrix Editor and
plotting windows of the TI-92.   In this activity you will examine some
examples of linear functions.

      Turn on, Press **MODE**, **F2**, select Split Screen → LEFT - RIGHT.   For
Split 1 Application select 6:Data/Matrix Editor and for Split 2 Application
select 4:Graph. Press **ENTER** twice to exit MODE.

      The TI-92 asks for a variable name.   Give a suitable name such as
exampl1.   Press **ENTER** twice and the screen will appear as two halves
known as windows. Figure 1.1 shows the screens for setting up the TI-92 as
described above.

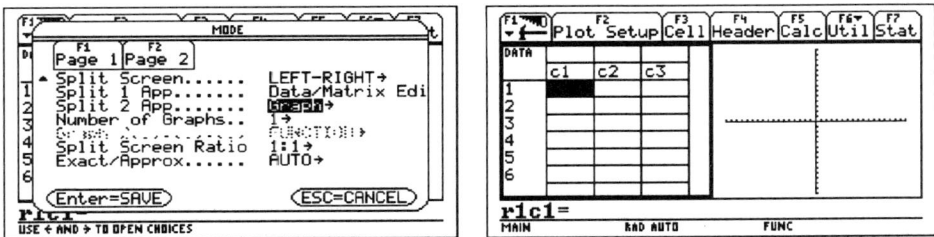

Figure 1.1   Setting up the TI-92 in split screen mode

(B)   To plot points you need to define a pair of coordinates (x, y). The x values give
the horizontal movement from the point at the centre of the screen where the
two axes cross and the y value gives the vertical movement.

      Press **2nd APPS** to highlight the GRAPH screen.   Press the cursor pad
and a cursor will appear on the GRAPH screen.   By continuously pressing the
cursor pad you can move the cursor to the right or left to increase or decrease

the x value and up and down to increase or decrease the y value. The (x,y) values are shown on the screen. Move the cursor to the following points (or as near as is possible)

$$(0,0) \ (2,1) \ (-1,1) \ (-1,-2) \text{ and } (0,-3)$$

(C)   Press **2nd APPS**. The left side of the screen is highlighted. This is taken up with the Data/Matrix Editor, which is used for inputting and managing data. You may be familiar with using tables to represent information (like a train timetable) - the Data/Matrix Editor is used in exactly the same way. The top left square should be highlighted, press **ENTER**.

"r1 c1" should appear on the input line. This refers to the part of the matrix which is in both row one and column one. We shall use column one for the x values and column two for the y values. Each x value having a corresponding y value which is shown in the same row.

Input the following points $(1,0)$, $(0,-1)$, $(-1,-2)$ and $(-2,-3)$ so that r1 c1 = 1, r1 c2 = 0, r2 c1 = 0, r2 c2 = -1, etc. We now want to plot these points on the graph. Press **2nd APPS** to highlight the GRAPH screen. Press ♦, **W** (i.e. choose Y =) and select Plot 1, **ENTER**. A table appears on the screen which asks for some information. Complete the table as in Figure 1.2.

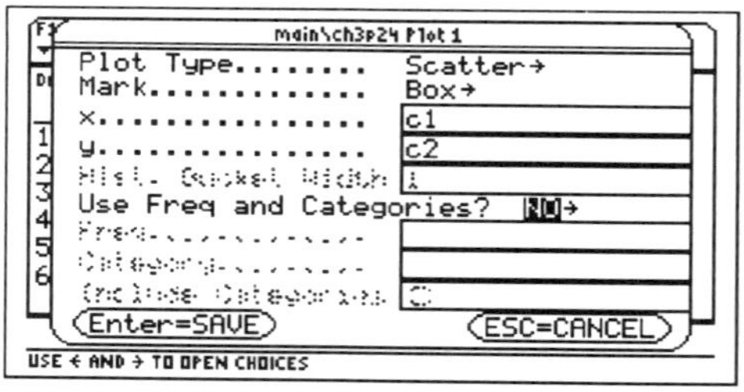

Figure 1.2   Entering a table of values

Press **ENTER** to Save, Press ♦ **R** (i.e. choose GRAPH) to return to the graph screen. Your graph should appear with the points plotted on the graph. What do you notice about these points?

Select the Y = Editor and enter $y1 = x - 1$, **ENTER** and GRAPH. You should see a straight line that passes through all the points that you plotted, see Figure 1.3.

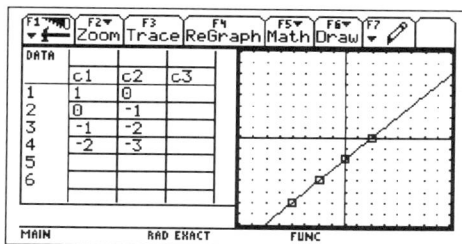

Figure 1.3   Drawing a graph through the data values

When the graph produced is a straight line, as in this example, we say that x and y are related by **a linear law**.

The equation $y = x - 1$ defines a relationship between the x and y coordinates.   The y coordinate is obtained by subtracting 1 from the x coordinate.

Clear the Y = Editor and Data/Matrix screen in the following way: in the Y = Editor screen, press **F1**, select 8:Clear Functions and **ENTER**. Switch to the Data/Matrix screen, press **F1**, select 8:Clear Editor and **ENTER**.

(D)   Plot the points (4,5), (–3,–2) and (0,1) using the Data/Matrix Editor.   In the Y = Editor define y1 = x + 1, **ENTER**.   Again a line should pass through the three points.

Repeat this activity for:

(a)   (2,5) (1,3) (–2,–3) and $y = 2x + 1$

(b)   (0,0) (2,4) (–1,–2) and $y = 2x$

(c)   (3,5) (1,1) (–2,–5) and $y = 2x - 1$

You should have 3 straight lines all of which are parallel.   What do the 3 equations have in common?

Where does each line cross the vertical axis?   What do you notice if you compare these values with the equations?

(E)   (i)   Clear the graph Y = Editor and Data/Matrix Editor.
Input the lines with equations

$$y1 = x + 1$$
$$y2 = 2x + 1$$
$$y3 = 3x + 1$$

(ii)   What do the 3 lines have in common? What do the 3 equations have in common?

(iii)   Which line is steepest?

In general the line with equation $y = mx + c$ has **gradient** m and crosses the vertical axis at c. The gradient tells you how steep the line is. The value c is often referred to as the **intercept**. Because the graph is a straight line we say that there is a linear law or relationship between x and y.

(F)   (i)   Clear the GRAPH screen and Data/Matrix editor. Plot each of the straight lines defined by the equations below.

$$y = -x + 2 \qquad y = \tfrac{1}{2}x + 1$$
$$y = -2x + 4 \qquad y = \tfrac{1}{4}x - 1$$
$$y = -3x - 2 \qquad y = \tfrac{1}{5}x + 2$$

(ii)   Find the gradient of each line. What does a negative gradient indicate? Also check that each line has the intercept that matches the equation.

1.2   THE EQUATION OF A STRAIGHT LINE

The equation of a straight line is given by $y = mx + c$ where m is the gradient of the line and c the intercept. The gradient gives the steepness of the line and is defined as

$$\text{gradient} = \frac{\text{change in } y}{\text{change in } x}$$

If the coordinates of two points on the line are (x_1, y_1) and (x_2, y_2) then the gradient is given by

$$\frac{y_2 - y_1}{x_2 - x_1}$$

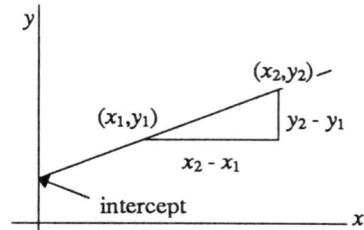

Figure 1.4

The intercept is the point where the line cuts the vertical or y-axis.

▓▓▓▓▓▓▓▓▓▓ **Example 1A** ▓▓▓▓▓▓▓▓▓▓▓▓▓▓▓▓▓▓▓▓▓▓▓▓▓▓▓▓

A straight line passes through the points (2,6) and (1,4).

(i)   Find the gradient of the line.

(ii)   Find the equation of the line.

Solution

(i)   The gradient m is given by

$$m = \frac{y_2 - y_1}{x_2 - x_1} = \frac{6-4}{2-1} = \frac{2}{1} = 2 \,.$$

(ii)   As the gradient of the line is 2 the equation must be of the form $y = 2x + c$. As the line passes through (1,4) we can replace y by 4 and x by 1 to give

$$4 = 2 \times 1 + c$$
$$4 = 2 + c$$

so $c = 2$ and the equation is $y = 2x + 2$.

▓▓▓▓▓▓▓▓▓▓ **Example 1B** ▓▓▓▓▓▓▓▓▓▓▓▓▓▓▓▓▓▓▓▓▓▓▓▓▓▓▓▓

Find any three points that lie on the line $y = 5x - 3$ and then draw the line.

Solution

Taking $x = 1$ gives $y = 5 \times 1 - 3 = 2$ so (1,2) is one point.
Taking $x = 2$ gives $y = 5 \times 2 - 3 = 7$ so (2,7) is a second point.
Taking $x = 0$ gives $y = 5 \times 0 - 3 = -3$ so (0,–3) is a third point.

Figure 1.5 shows these 3 points and a line drawn through them.

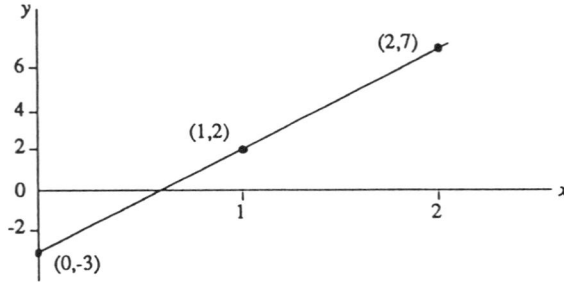

Figure 1.5

1. For each equation given below find 3 points that lie on the line and then draw the line that passes through them.

    (a) $y = 2x + 3$ (e) $y = -2x + 1$
    (b) $y = x - 4$ (f) $y = 6x - 2$
    (c) $y = 5x - 4$ (g) $y = -4x + 6$
    (d) $y = \frac{1}{2}x + 1$ (h) $y = -2x - 4$

2. Find the gradient and the equation of the line that passes through each pair of points given below.

    (a) (0,0) (2,6)   (e) (3,0) (6,1)
    (b) (1,1) (4,5)   (f) (5,4) (0,5)
    (c) (3,7) (0,4)   (g) (1,4) (3,–2)
    (d) (0,6) (2,2)   (h) (–1,6) (5,–2)

1.3   PROPORTIONALITY

If a linear equation is of the form

$$y = kx$$

then y and x are said to be **proportional** and in symbols this is often written as $y \propto x$ (y is proportional to x). The constant k is known as the **constant of proportionality**. If y is proportional to x then the graph of the equation $y = kx$ is a straight line with gradient k passing through the origin.

Table 1.1 below gives values for the tension in a spring and its extension.

Table 1.1

Tension T (N)	0	0.5	1.0	1.5	2
Extension e (cm)	0	4	8	12	16

Show that the tension is proportional to the extension and find the constant of proportionality.

Solution

Figure 1.6 shows a graph of tension against extension.

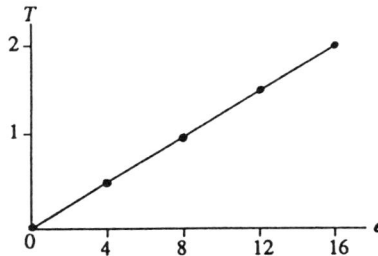

Figure 1.6

As we have a straight line passing through (0,0) then the two quantities must be proportional.

As $T \propto e$, then $T = ke$. When the tension is 16 the extension is 2, so $16 = k \times 2$, and $k = 8$.

▓▓▓▓▓▓▓▓▓▓▓ **Example 1D** ▓▓▓▓▓▓▓▓▓▓▓▓▓▓▓▓▓▓▓▓▓▓▓

The speed at which a stone hits the ground is proportional to the square root of the height from which it was released. A stone dropped from a height of 4 m hits the ground at 9 ms^{-1}. Find the constant of proportionality. Calculate the speed of the stone when it is dropped from a height of 8 m.

Solution

Using v for speed and h for height gives

$$v \propto \sqrt{h}$$

or

$$v = k\sqrt{h}$$

Substituting in the known values for v and h gives

$$9 = k \times \sqrt{4}$$

Solving for k we have

$$k = 4.5$$

Thus $v = 4.5\sqrt{h}$

When $h = 8$ the speed is given by

$$v = 4.5\sqrt{8}$$
$$= 12.73 \text{ ms}^{-1}$$

Exercise 1B

1.   For each set of data given below, check to see if the two variables are proportional. If they are find the constant of proportionality.

(a)

P	0	1	3	5
Q	1	2	4	7

(c)

x	1	2	5	9
y	2	3	6	10

(b)

R	1	2.5	3	5
S	10	25	30	50

(d)

v	1.1	1.2	1.3	1.5	2
t	1.65	1.80	1.95	2.25	3.00

2.   The tension in a spring is proportional to its extension. A spring stretches 8 cm under a tension of 5 N. Find the constant of proportionality. What tension would be present if the extension were 10 cm?

3.   The petrol consumed by a car is proportional to the distance travelled. If 300 miles can be travelled using 60 litres of petrol, find the constant of proportionality. How far could the car travel on 100 litres of petrol?

4.   The diameter of the stem of a plant is proportional to the square root of its growing time. A plant that has been growing for 36 days has a diameter of 0.9 cm. What will the diameter be after 100 days?

5.   The friction force F acting on a sliding block is proportional to the normal reaction force R. Complete Table 1.2 and find the equation relating F and R.

Table 1.2

Normal Reaction R(N)	19.8	16.2	
Friction F(N)	6.1		4.7

6.   The pressure P of a quantity of gas is proportional to its temperature, T and is proportional to its density, ρ. Write down an equation describing the law between P, T and ρ.

1.4   MODELLING WITH LINEAR LAWS

Often data collected from experiments lie close to a straight line which does not pass through the origin. For example Figure 1.7 shows a graph of the length ℓ of a heated copper bar (in centimetres) against the temperature T of the bar (in degrees celsius).

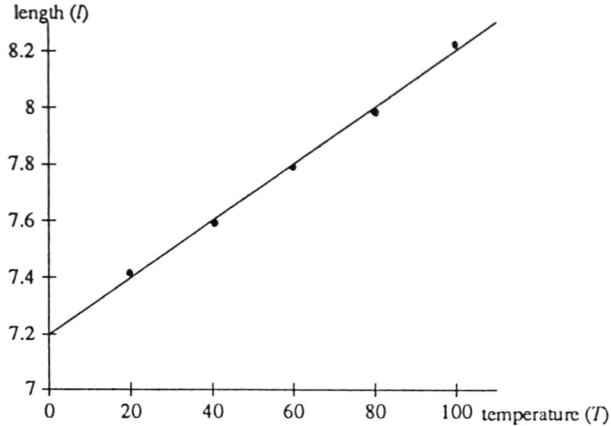

Figure 1.7

The data points are very close to a straight line called the 'line of best fit'. A simple rule for drawing this line is to make sure that there are as many points on one side of the line as the other.

The equation of the straight line is found by calculating its gradient and intercept

$$\ell = 0.01T + 7.2$$

We say that there is a **linear relationship** or a **linear model** between the variables ℓ and T. The most commonly occurring models in science and engineering are linear models.

The following example shows how we use the TI-92 to find linear models between variables.

Example 1E

In an experiment to investigate the tension in an elastic string the tension in the string and length of the string were measured and the results shown in Table 1.3.

Table 1.3   Experimental data investigating friction

length of spring ℓ (in centimetres)	20.0	38.5	48.0	57.0	67.0
tension in spring T (in newtons)	10.0	17.5	21.5	25.5	29.0

Show that a linear law is a good model relating T and ℓ and find the equation of the 'line of best fit'.

Solution

The first step is to plot the graph of the data. Reading the data values in pairs, the graph suggests that ℓ and T are linearly related.

Turn on the TI-92. Split the screen with Data/Matrix Editor in the left window and GRAPH in the right window. Input the data into the matrix editor with length of spring in Column 1 and Tension in spring in Column 2. With Data/Matrix screen highlighted

Press select **F5** 5:LinReg

Figure 1.8 shows the screen for this activity.

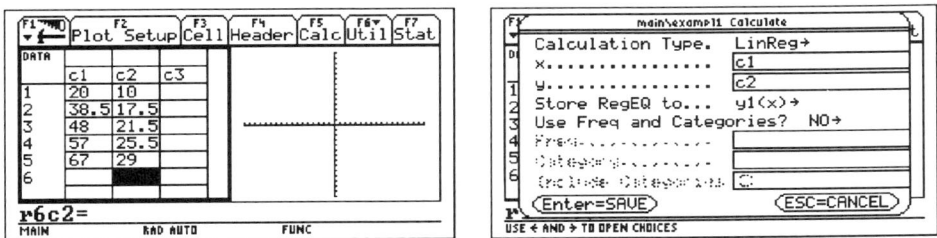

Figure 1.8

Press **ENTER** and STAT VARS is displayed. This gives a line of best fit of the form $y = ax + b$, this line has also been stored in y1(x) of the Y = Editor.

Press **2nd APPS**

Press Y = and check that Plot 1 and y1 = are selected. To display the graphs we need
to choose an appropriate sized window. Press WINDOW and type in the following:

xmin = –10, xmax = 70, xscl = 5
ymin = –10, ymax = 30, yscl = 5

Now select GRAPH.

The graph should then appear with the data points and the line of best fit
running through them. The line of best fit coefficients and graphs are shown in
Figure 1.9.

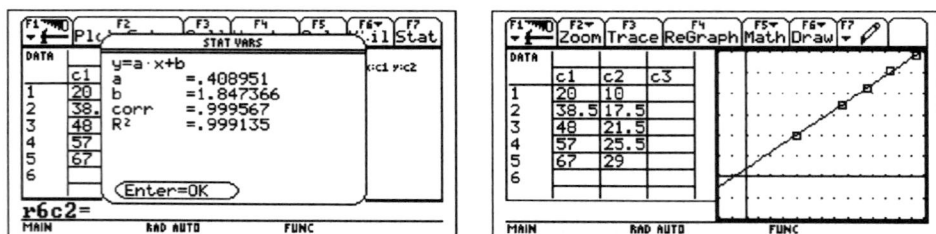

Figure 1.9   Linear model relating ℓ and T

The linear model relating T and ℓ is approximately given by

$$T = 0.41\,\ell + 1.85$$

In this example we used the LinReg calculation which produces a line of best
fit using a method called Linear Regression. This method minimises the sum of the
squares of the distances between each point and the straight line. The technique is
often called the method of least squares.

Exercise 1C

1.   Use the TI-92 to find the equation of the line of best fit for the following sets of data.

(a)
x	0.1	0.5	0.7	1.1
y	1.1	2.6	3.6	5.1

(b)
r	0.1	0.5	1.0	1.4	1.7	2.1
s	−3.89	−2.65	−1.1	0.14	1.07	2.31

(c)
t	1	2	3	4
v	10.2	0.4	−9.4	−19.2

(d)
x	1.1	1.3	1.5	1.7	1.9
p	2.20	2.55	2.89	3.23	3.58

2.   In an experiment the resistance of a length of wire was measured at different temperatures and the readings are shown in Table 1.4.

Table 1.4

Temperature T (°C)	50	80	100	120	160
Resistance R (ohms)	53.3	58.4	61.9	65.3	72.3

(a)   Draw a graph of the resistance against temperature and show that a linear law is a good model.

(b)   Find the equation of the line of best fit.

(c)   Use the model in part (b) to find the resistance of the wire at a temperature of 20°C.

3.   The speed of a car (in ms^{-1}) accelerating away from traffic lights is given at 0.5 second time intervals in Table 1.5. From a linear model find the acceleration of the car.

Table 1.5

time t (seconds)	0.5	1.0	1.5	2.0	2.5	3.0
speed v (ms^{-1})	1.58	3.26	4.84	6.38	8.24	9.72

1.5   SOLVING LINEAR EQUATIONS

In this section we review the procedure for solving linear equations. Consider the problem of solving the linear equation

$$4x - 2 = 10$$

The first step is to add 2 to each side of the equation

$$4x - 2 + 2 = 10 + 2$$

$$4x = 12$$

Now dividing both sides by 4 gives

$$x = \frac{12}{4} = 3$$

Sometimes the unknown variable x will appear on both sides of the equals sign, as in

$$5x + 12 = 7x + 4$$

The procedure is to collect the x's on one side of the equation and the numbers on the other side. Subtract $5x$ from both sides gives

$$12 = 2x + 4$$

Now subtract 4 from each side.

$$8 = 2x$$

and dividing both sides by 2 gives the answer

$$4 = x$$

(or more usually $x = 4$).

TI-92 ACTIVITY 1B

In this Activity we show how the TI-92 can be used to solve linear equations directly. Clear the HOME screen.

(A)   Consider again the problem

$$5x + 12 = 7x + 4$$

Press **F2** and select 1:solve(

Type $5x + 12 = 7x + 4, x$) **ENTER**

The TI-92 gives the answer $x = 4$. The parameters for the solve command are

solve(equation, variable)

Figure 1.10 shows the HOME screen for solving several linear equations.

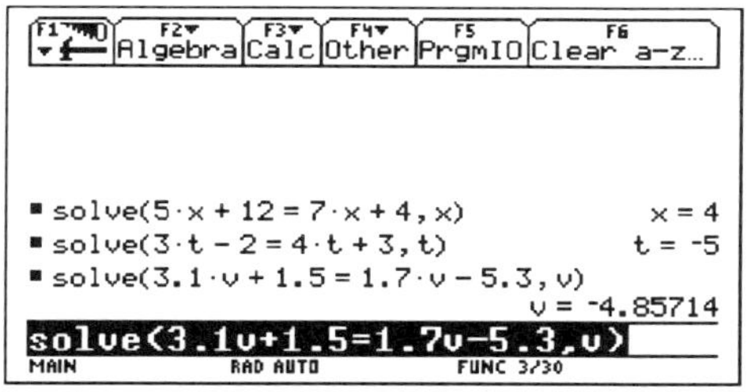

Figure 1.10   Solving linear equations on the TI-92

(B)   Use the TI-92 to solve the following equations for the unknown variable.

(a) $3(2x - 1) = 24$ (b) $4x - 2 = 5x + 1$

(c) $4y - 1 = 11$ (d) $3u + 2 = 2(7 - u)$

(e) $13.03t + 4.82 = -0.61t - 2.17$ (f) $4.21x - 1.73 = 2.63x + 5.19$

Forming and Solving Equations

░░░░░░░░░░░ **Example 1F** ░░░░░░░░░░░░░░░░░░░░░░░░░░░░

When three consecutive numbers are added together their total is 144. What are these numbers?

Solution

Let x be the smallest of the three numbers, then $x + 1$ and $x + 2$ will be the other two. So their total is

$$x + x + 1 + x + 2 = 144$$

or $3x + 3 = 144$

To solve this subtract 3 from both sides of the equation to give

$$3x = 141$$

and then divide by 3 to give

$$x = \frac{141}{3} = 47$$

So the lowest number is 47 and the others are 48 and 49.

░░░░░░░░░░░ **Example 1G** ░░░░░░░░░░░░░░░░░░░░░░░░░░░░

The perimeter of a triangle is 27 cm. The longest side is 3 times as long as the shortest side and the other side is 8 cm shorter than the longest side. Find the length of each side.

Solution

Let x be the length of the shortest side, then the other two sides have lengths $3x$ and $3x - 8$. So the perimeter is given by

$$x + 3x + 3x - 8 = 27$$

or $7x - 8 = 27$

Adding 8 to both sides gives

$7x = 35$

Dividing by 7 gives

$x = 5$ cm

So the sides are 5 cm, 15 cm and 7 cm.

Exercise 1D

1. For each of the following problems first solve the equation 'by hand' and then check your answer with the TI-92.

(a)	$6x - 18 = 12$	(g)	$16 - 2x = x + 10$
(b)	$4x + 2 = 6x - 9$	(h)	$4x + 2 = 8 - x$
(c)	$7x + 5 = 11$	(i)	$7t - 11 = 2t + 9$
(d)	$6 - 3x = 4 + 2x$	(j)	$3(t-4) = 2(t+7)$
(e)	$16x - 12 = 4$	(k)	$0.21(x+1.2) = 0.37x$
(f)	$5x + 2 = -13$	(l)	$1.46(x-0.9) = -4.1(2.61-2x)$

2. The sum of 3 consecutive odd numbers is 141. What are the numbers?

3. The perimeter of a rectangle is 60 cm. Find the length and width if

   (a) the length is twice the width,
   (b) the length is 8 cm greater than the width.

4. Using the formula $s = ut + \frac{1}{2}at^2$ find the unknown term if

   (a) $t = 1$, $s = 6$, $u = 5$,
   (b) $t = 2$, $s = 4$, $a = -2$,
   (c) $t = 10$, $s = 5$, $a = 0.5$.

1.6   SIMULTANEOUS EQUATIONS

Often two equations may apply at the same time, for example $y = 4x + 6$ and $y = 3x + 10$. These are known as **simultaneous equations**. Is there a pair of values for x and y which will satisfy both equations? If so how do we find them?

TI-92 ACTIVITY 1C

The aim of this Activity is to use the TI-92 to solve simultaneous equations. Clear the HOME screen and the Y = Editor.

(A)   Suppose that we have the two equations above.
      In the Y = Editor define $y1 = 4x + 6$ and $y2 = 3x + 10$. GRAPH each equation. By pressing the cursor pad move the cursor to the point where the lines cross. This is the point of intersection and should have an x value of about 4 and y value of about 22. Return to HOME screen.
      We know that $y = 4x + 6$ and $y = 3x + 10$. Therefore

$$4x + 4 = 3x + 10$$

Press **F2**, select 1:solve, **ENTER**. Type $4x + 6 = 3x + 10,x$) **ENTER**. This gives $x = 4$. Substituting this answer into either of the original equations gives the y-value.

(B)   Use the approach above to solve the following pairs of equations; adjusting the scales of your plot as necessary.

   (i) $y = 20 - 5x$ (iii) $y = 10 - x$
          $y = 4 + 3x$ $y = 4 + 3x$

   (ii) $y = 2x + 4$ (iv) $y = 7 - 2x$
          $y = x + 10$ $y = 4 + 3x$

(C)   (i)   Often simultaneous equations are presented in the form $4x + 2y = 8$ and $6x - 5y = 10$.
            Plot the first of these equations. Now plot the other equation and with the cursor try to read off the values of x and y where the lines intersect. In the HOME screen solve the equations with respect to y i.e. solve $(4x + 2y = 8,y)$ and solve $(6x - 5y = 10,y)$.
            Now solve these rearranged equations as before in part (B). These two methods of solution are shown in the screen dumps in Figure 1.11.

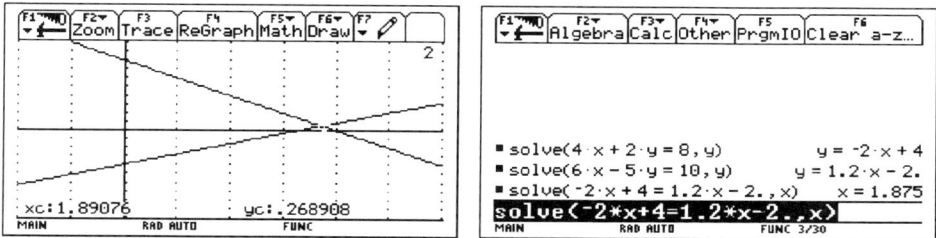

Figure 1.11

(D)   Use the approach of (C) to solve the simultaneous equations below.

(i) $x + y = 6$ (iii) $2x + 3y = 5$
        $x - y = 7$ $3x - 2y = 4$

(ii) $2x + 3y = 6$ (iv) $x + y = 6$
        $5x - 2y = 10$ $2x - y = 7$

(E)   Finally try the following.
       Consider

$$4x + 2y = \ 8$$
$$6x - 5y = 10$$

First solve (4x + 2y = 8,x) to give $x = -\dfrac{(y-4)}{2}$.

Now solve $(6x - 5y = 10,y) \,|\, x = -(y-4)/2$.
This expression means: solve 6x − 5y = 10 for y 'with'

$$x = -\frac{(y-4)}{2}$$

(The vertical bar is obtained using **2nd K** and is called the 'with' operator.)

The TI-92 gives $y = \frac{1}{4}$. To obtain x

Solve $(x = -(y-4)/2,x) \,|\, y = 1/4$

The solution is $x = 15/8$. Figure 1.12 shows the HOME screen for the activity.

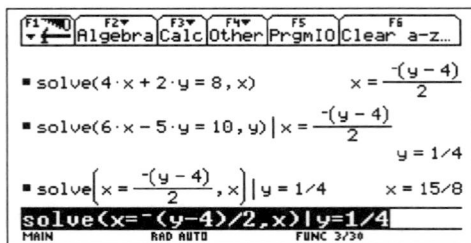

Figure 1.12

Pairs of simultaneous equations can be solved using the substitution approach described. However, with the TI-92 the pair of equations

$$ax + by = c \quad \text{and} \quad dx + ey = f$$

can be solved directly by using the approaches in section (E) and **Solve**.

(F)   Solve the simultaneous equations in (D) directly by using the TI-92.

Solving Simultaneous Equations

The method of solving simultaneous equations in TI-92 Activity 1C was to substitute the formula for y from one equation into the other. This method works well but can involve some tedious algebra. An alternative approach when solving simultaneous equations by hand is to add multiples of the two equations together to eliminate one variable. This is illustrated in the examples below.

Example 1H

Solve the pair of simultaneous equations

$$3x + 2y = 6$$
$$x + 2y = 4$$

Solution

First it is helpful to label the equations (1) and (2) as below.

$$3x + 2y = 6 \qquad (1)$$
$$x + 2y = 4 \qquad (2)$$

Here both equations contain the same number of y's so equation (2) can be subtracted from equation (1) to give

$$(3x - x) + (2y - 2y) = 6 - 4$$
$$2x \qquad\qquad = 2$$

So $x = 1$. Substituting this into the first of the original equations gives

$$3 + 2y = 6$$
$$2y = 3$$
$$y = 1.5.$$

So the solution is $x = 1$ and $y = 1.5$.

░░░░░░░░░░░ **Example 1I** ░░░░░░░░░░░░░░░░░░░░░░░░░░░

Solve the pair of simultaneous equations

$$3x - 4y = 2$$
$$2x + 3y = 6$$

Solution

Labelling the equations (1) and (2) is helpful,

$$3x - 4y = 2 \qquad (1)$$
$$2x + 3y = 7 \qquad (2)$$

As a first step multiply equation (1) by 2 and equation (2) by 3. This will mean that both equations end up with the same number of x's in each.

$$6x - 8y = \;\; 4 \qquad 2 \times (1)$$
$$6x + 9y = 21 \qquad 3 \times (2)$$

Now subtracting the second of these from the first gives

$$(6x - 6x) + (-8y - 9y) = 4 - 21$$
$$-17y = -17$$
$$y = 1$$

Substituting this value of y into the first equation and solving gives $x = 2$. So the solution is $x = 2$ and $y = 1$.

░░░░░░░░░░░░░░ **Exercise 1E** ░░░░░░░░░░░░░░░░░░░░░░░░░░░░░░

Solve each of the following pairs of simultaneous equations 'by hand'.

(a) $x + y = 16$
      $x - y = 4$

(d) $7x - 2y = 18$
      $3x + 4y = 2$

(b) $2x + 3y = 12$
      $4x + 5y = 22$

(e) $4a + 5b = 6$
      $3a - 2b = 16$

(c) $4x + 3y = 30$
      $5x - 2y = 26$

(f) $3u + 6t = 24$
      $4u + 5t = 17$

(Check your answers using the TI-92).

░░░

1.7   QUADRATICS

In this section we introduce another rule between two variables x and y which uses the term x^2 as well as x. The rule is called a **quadratic law** and has the general form $y = ax^2 + bx + c$. For example $y = x^2 + 2x - 3$ is a quadratic law.

TI-92 ACTIVITY 1D

Clear the HOME screen, the Data/Matrix Editor and Y = Editor.

(A)   Many physical quantities are related by quadratic laws. As a stone falls the distance that it has travelled after certain times are given in the Table 1.6.

Table 1.6

time (seconds)	0	1	2	3	4
distance fallen (m)	0	5	20	45	80

(i)   Split the screen with the left window as the Data/Matrix Editor and the right window as the GRAPH screen. Enter the data in Table 1.6 with time in column 1 and distance in column 2.

      In the GRAPH screen, set the scale and range (for the window) to:

$$xmin = -10, xmax = 10, xscl = 2$$
$$ymin = -10, ymax = 100, yscl = 10$$

Return to graph screen using GRAPH.

(ii)   Return to the Data/Matrix Editor. It is possible to find the equation of a quadratic curve that passes through the points.

In the Data/Matrix Editor window Press **F5** and complete the calculation screen shown in Figure 1.13.

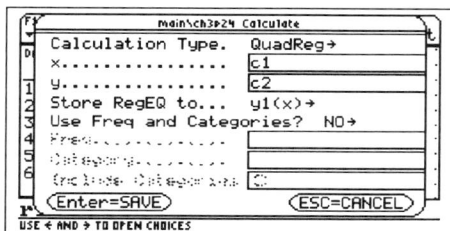

Figure 1.13	Figure 1.14
The QuadReg command	Quadratic graph of best fit

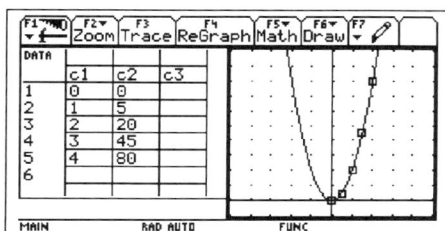

Choose GRAPH and a quadratic graph of best fit is shown through the points. The screen dump for these activities are shown in Figure 1.14.

The quadratic curve of Best Fit is given approximately by $y = 5x^2$ and this is a quadratic model for the data of Table 1.6.

(B)   The Highway Code gives the following table of speed and stopping distances.

Table 1.7

Speed (ms^{-1})	8.9	13.3	17.8	22.2	26.7	31.1
Stopping Distance (m)	12	23	36	53	73	96

Enter the data into the Data/Matrix Editor and plot the points. Then use the QuadReg command to find the equation of the quadratic curve through these points, plotting the result.

(Note that in the Highway Code the speeds are given in mph, they have been converted to ms^{-1} using the rule 1 mph $\equiv$ 0.444 ms^{-1}).

(C) (i) Close the Data/Matrix Editor window and Clear the Y = Editor.
Set WINDOW to

$$xmin = -4, xmax = 4, xscl = 1,$$
$$ymin = -4, ymax = 4, yscl = 1$$

Graph the expressions.

(a) $y = x^2$ (c) $y = x^2 + 2$

(b) $y = x^2 + 1$ (d) $y = x^2 - 1$

Describe how each curve is related to $y = x^2$.

(ii) Clear the Y = Editor. Change the WINDOW so that the graphs of each
of the expressions below are shown on the screen.

(a) $y = x^2$ (c) $y = (x - 2)^2$

(b) $y = (x - 1)^2$ (d) $y = (x + 2)^2$

Describe how each curve is related to $y = x^2$.

(iii) For each of the following expressions try to predict the coordinates of
the lowest point of the curve. Graph to check your predictions.

(a) $y = x^2 + 1$ (d) $y = (x + 2)^2 + 4$

(b) $y = (x + 1)^2$ (e) $y = -x^2$

(c) $y = (x + 1)^2 - 1$ (f) $y = 4 - x^2$

This activity (C) shows that all quadratics are similar in shape to the basic
quadratic $y = x^2$. The effect of the 'other numbers' is to translate the basic shape of the
graph of $y = x^2$.

The curve $y = x^2 + a$ is translated up a units (if a is positive) and has a
minimum value of a when $x = 0$.

The curve $y = (x-b)^2$ is translated to the right by b units (if b is positive) and
has a minimum value of zero when $x = b$.

The curve $y = (x-b)^2 + a$ is translated to the right by b units and up by a units
and has a minimum value of a when $x = b$.

Solving Quadratic Equations

A quadratic equation is an equation of the form

$$x^2 + 5x + 6 = 0$$

The approaches that you have used for solving linear equations are not appropriate here. The TI-92 Activity 1E leads you into some of the important ideas needed for solving quadratic equations.

TI-92 ACTIVITY 1E

Clear the HOME screen, Y = Editor and GRAPH screen.

(A)   (i)   Graph $x^2 + 5x + 6$. You will see that the curve crosses the x-axis twice and so has 2 solutions. Return to the HOME screen and use **F2** 1:solve to find the value of these 2 solutions.

       (ii)   Repeat for $x^2 + 6x + 9$. This time you will see that there is only one solution.

      (iii)   Repeat for $x^2 + 5x + 10$. This time you will see that the curve does not cross the axis and there are no real solutions. Using solve will give an error message "false". Now try using cSolve (Complex Solve found using **F2 A**); this will produce strange looking expressions involving î. These are known as **complex numbers** and will be dealt with in Chapter 10.

(B)   (i)   Clear the HOME screen. Type

           factor($x^2 + 5x + 6$,x)

        What is the connection between the new expression and the solutions.

       (ii)   Repeat for $x^2 + 6x + 9$.

      (iii)   What happens for $x^2 + 5x + 10$?

(C)   (i)   Clear the HOME screen. Type

           expand((x + a)(x + b),x)

        Explain how $(x+a)(x+b)$ is related to $x^2 + px + q$.
        Write down p and q in terms of a and b.

(ii)   Write down what you would expect to get when you expand each of the expressions below and then check your answer on the TI-92.

     (a) $(x+2)(x+4)$
     (b) $(x+1)(x+5)$
     (c) $(x+3)(x-4)$
     (d) $(x-6)(x-2)$

Now graph each expression and find the solutions of the quadratic equation formed when each expression is put equal to zero.

(iii)   When $x^2 + px + q$ is factored it produces the two brackets $(x+a)(x+b)$ where $ab = q$ and $a + b = p$. For example with

$$x^2 + 7x + 10$$

we have $ab = 10$ and $a + b = 7$. Clearly these are satisfied by $a = 5$ and $b = 2$, so

$$x^2 + 7x + 10 = (x+5)(x+2)$$

Try to factorise each expression below, checking your answers with the TI-92.

(a)	$x^2 + 5x + 4$	(d)	$x^2 - 5x + 6$
(b)	$x^2 + 6x + 8$	(e)	$x^2 - x - 6$
(c)	$x^2 + 3x + 2$	(f)	$x^2 + x - 6$

A quadratic equation can have

(i)   two roots if it crosses the x-axis twice,
(ii)   one root if it just touches the x-axis,
(iii)   two complex roots if it does not cross the x-axis.

The quadratic equation $x^2 + px + q$ has factors $(x+a)$ and $(x+b)$ where $a + b = p$ and $ab = q$.

Solving Quadratics by Factorisation

It is possible to solve quadratics by factorisation, an approach which is demonstrated in the examples below.

▓▓▓▓▓▓▓▓▓▓▓▓▓ **Example 1J** ▓▓▓▓▓▓▓▓▓▓▓▓▓▓▓▓▓▓▓▓▓▓▓▓▓

Solve $x^2 + 7x + 12 = 0$ by factorisation.

Solution

We have to write $x^2 + 7x + 12 = 0$ in the form $(x+a)(x+b) = 0$.

So $a + b = 7$ and $ab = 12$. Possible values of a and b that have a product of 12 are 1 and 12, 2 and 6, 3 and 4. But a and b must also have a sum of 7 and so $a = 3$ and $b = 4$, giving

$$(x+3)(x+4) = 0$$

The product of these brackets will be zero when either bracket is zero i.e. when $x + 3 = 0$ or $x + 4 = 0$ giving $x = -3$ and $x = -4$ as the solutions of the equation.

▓▓▓▓▓▓▓▓▓▓▓▓▓ **Example 1K** ▓▓▓▓▓▓▓▓▓▓▓▓▓▓▓▓▓▓▓▓▓▓▓▓▓

Solve $3x^2 + 14x + 8 = 0$ by factorisation.

Solution

The equation $3x^2 + 14x + 8 = 0$ must be written in the form $(3x+a)(x+b) = 0$. So $ab = 8$ and $a + 3b = 14$. Possible values of a and b that have a product of 8 are 1 and 8 or 2 and 4. To satisfy $a + 3b = 14$ we must use $a = 2$ and $b = 4$ giving

$$(3x+2)(x+4) = 0$$

The product of these brackets will be zero when either is zero giving

$$3x + 2 = 0 \quad \text{or} \quad x + 4 = 0$$
$$x = -\tfrac{2}{3} \quad \text{or} \quad x = -4$$

So the solutions of the equation are $x = -\tfrac{2}{3}$ and $x = -4$.

Solving Quadratics with the Formula

Often quadratic equations are difficult or impossible to solve using the technique of factorisation. However there is a formula to solve the quadratic equation $ax^2 + bx + c = 0$. It is given by

$$x = \frac{-b \pm \sqrt{b^2 - 4ac}}{2a}$$

When using this formula, note that

(i)   if $b^2 - 4ac > 0$ then there will be 2 solutions,

(ii)   if $b^2 - 4ac = 0$ then there will be only 1 solution,

(iii)   if $b^2 - 4ac < 0$ there will be no real solutions as the square root of a negative number is not defined (until Chapter 10).

[We will not prove the formula here. A derivation of the formula is shown on page 344 of the TI-92 User Manual.]

Example 1L

Find the solutions of

$$3x^2 + 4x - 8 = 0$$

Solution

Here

$$a = 3, \quad b = 4 \quad \text{and} \quad c = -8.$$

Substituting into the formula gives

$$x = \frac{-4 \pm \sqrt{4^2 - 4 \times 3 \times (-8)}}{2 \times 3} = \frac{-4 \pm \sqrt{16 + 96}}{6} = \frac{-4 \pm \sqrt{112}}{6}$$

Hence $x = 1.10$ or -2.43 (to 3sf).

Exercise 1F

In each of the following problems, do the activity first 'by hand' and then check your answers with the TI-92.

1.   Sketch the following quadratics.

   (a) $y = x^2 + 6$ (d) $y = 2 - (x+3)^2$

   (b) $y = (x+4)^2$ (e) $y = (x+3)^2$

   (c) $y = 6 - x^2$ (f) $y = (x-4)^2 + 2$

2.   Factorise the following quadratics.

   (a) $x^2 + 9x + 20$ (f) $2x^2 - 9x - 5$

   (b) $x^2 - 7x + 12$ (g) $2x^2 + 5x + 2$

   (c) $x^2 + 4x + 3$ (h) $5x^2 - 34x - 7$

   (d) $x^2 + 5x + 6$ (i) $3x^2 - 10x - 8$

   (e) $x^2 - 1$ (j) $x^2 - 4$

3.   Solve the following quadratic equations by factorisation.

   (a) $x^2 + 6x + 9 = 0$ (c) $2x^2 + x - 10 = 0$

   (b) $x^2 + 10x + 16 = 0$ (d) $x^2 - 16 = 0$

4.   Find the solutions, if any, of the following quadratic equations.

   (a) $x^2 + 6x - 7 = 0$ (e) $6x^2 - 10x + 2 = 0$

   (b) $3x^2 + 4x - 9 = 0$ (f) $5x^2 - 9x + 8 = 0$

   (c) $x^2 + 8x + 20 = 0$ (g) $13x^2 + 81x - 94 = 0$

   (d) $3x^2 - 4.5x + 9 = 0$ (h) $6x^2 - 42 = 0$

1.8   POLYNOMIALS

The linear and quadratic functions considered so far in this chapter are simple examples of **polynomials**. A polynomial is any function of the form

$$a_0 + a_1 x + a_2 x^2 + a_3 x^3 + ... + a_n x^n$$

The numbers a_0, a_1, a_2, ..., a_n are referred to as the **coefficients** of the polynomial and n the largest power is called the **degree** of the polynomial.

TI-92 ACTIVITY 1F

Clear the Y = Editor.

(A)   Graph the following functions using the TI-92.

   (i) x^2 (ii) x^3 (iii) x^4 (iv) x^5 (v) x^6 (vi) x^7 (vii) $-x^3$ (viii) $-x^4$

   What do you notice? Clear all the graphs.

(B)   Graph the following.

   (i) x^3 (ii) $x^3 + 2$ (iii) $x^3 - x$ (iv) $x^3 - x^2$

   How many solutions can a polynomial of degree 3 have? Is there a minimum number of solutions?

(C)   Investigate how many times a polynomial of degree 4 crosses the x-axis.

   By choosing polynomials of different degree show that the number of roots (and factors) of a polynomial is less than or equal to the degree.

(D)   It is possible to express polynomials in the form

$$(x-a)(x-b)(x-c) \ .....$$

   Try graphing this expression for different numbers of brackets and different values in the brackets. What do you notice? What happens if some brackets are repeated?

░░░░░░░░░░░░░░░░ **Exercise 1G** ░░░░░░░░░░░░░░░░░░░░░░░░░░░░░░░░░░

Sketch each of the following polynomials.

(a) $y = x^5 + 1$ (b) $y = x^6 - 1$ (c) $y = (x+1)^4 - 1$

(d) $y = (x+1)(x-1)(x+2)$ (e) $y = (x+1)^2(x-1)$ (f) $y = x(x-2)^2$

1.9 FUNCTIONS

In earlier sections we have introduced the idea of variables being related to each other by rules or laws, for example, the tension in an elastic string T is linearly related to its length, ℓ. The measured quantities T and ℓ are called **variables** because their values vary in an experiment to find the linear law. In such an experiment we would probably choose the tension and then measure the corresponding length. The chosen variable is called the **independent variable** and the other variable is called the **dependent variable** because its value depends on the value chosen for the independent variable.

When the two variables x and y say are related so that the one quantity y depends on the value of the other x, then y is said to be a **function** of x. The only restriction on such a relationship being called a function is that for each possible value of x there is only one value of y. An example of a quadratic function is

$$y = x^2 + 6 \quad \text{for} \quad x \geq 0.$$

Notice that the independent variable is restricted to positive values of x. This is called the **domain** of the function. The values of y are restricted to $y \geq 6$ and this set of values is called the **range** of the function.

It is often convenient to use a special notation for functions. For example, instead of writing

$$y = x^2 + 6$$

we could write

$$f(x) = x^2 + 6$$

The value of a function for a particular value of x is written in a special way. For example, suppose that we want the value of $f(x)$ for $x = 5$. Then we denote this by $f(5) = 5^2 + 6 = 31$. Similarly $f(2) = 2^2 + 6 = 10$ is the value of $f(x)$ for $x = 2$.

Different letters can be used to define different functions, for example

$$g(x) = \frac{x}{2} \quad \text{and} \quad b(x) = x + 5$$

Consider $f(x) = 2x + 1$ for the domain $-4 \le x \le 4$. A graph of this function shows that to each value of x there is only one value of y in the range $-7 \le y \le 9$. This is an example of a **one-to-one mapping** since for each y-value in the range there is a unique x-value in the domain (see Figure 1.15).

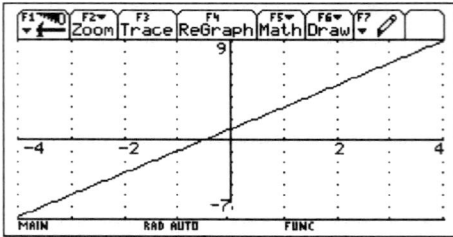

Figure 1.15   Graph of $y = 2x + 1$ Figure 1.16   Graph of $y = x^2 + 2x - 3$

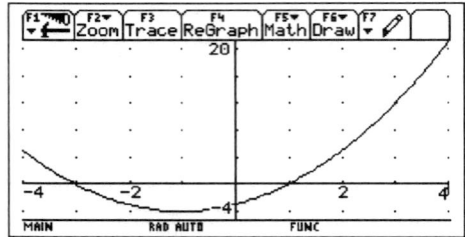

The graph of the function $f(x) = x^2 + 2x - 3$ in Figure 1.16 shows a different behaviour. For the domain $-4 \le x \le 4$ the range is $-4 \le y \le 21$, however each y-value in the range does not have a unique corresponding x-value in the domain. This is called a **many-to-one mapping**.

TI-92 ACTIVITY 1G

(A)   It is a simple matter to define functions with the TI-92. User defined functions can be a great time-saver when you need to repeat the same expression (but with different values). User defined functions can also extend your TI-92's capabilities beyond the built-in functions.

(i)   Suppose we want to define the function $f(x) = 5x$. This can be done in different ways using the STO▷ command, Define command or by the Program Editor.

In programming, multi-statement functions such as IF, ELSEIF, RETURN etc. are extremely powerful. So we shall consider the Define command which is the most useful of the ways of predefining a function.

Press **F4**, select 1:Define. Press **ENTER**. Type f(x) = 5x. Press **ENTER**.

The TI-92 then assigns the rule

f(x) = 5x

Type f(5) and press **ENTER**.

The TI-92 evaluates $f(5) = 5 \times 5 = 25$.

(ii)   Repeat the process to define $g(x) = x + 2$ and $h(x) = x/2$.
Find $h(2)$, $g(7)$ and $f(3)$.
Figure 1.17 shows the HOME screen for this activity.

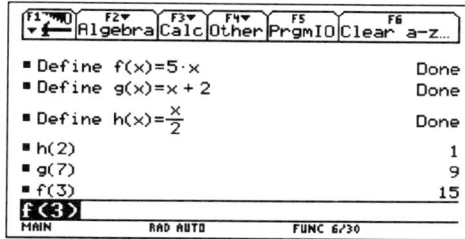

```
┌──────────────────────────────────────────────────────┐
│ F1▼  F2▼   F3▼   F4▼   F5        F6                    │
│ ▼€─Algebra Calc Other PrgmIO Clear a-z…                │
├──────────────────────────────────────────────────────┤
│ ▪ Define f(x)=5·x                          Done        │
│ ▪ Define g(x)=x + 2                        Done        │
│                    x                                    │
│ ▪ Define h(x)=─                            Done        │
│                    2                                    │
│ ▪ h(2)                                        1        │
│ ▪ g(7)                                        9        │
│ ▪ f(3)                                       15        │
│ f(3)                                                   │
│ MAIN         RAD AUTO        FUNC 6/30                 │
└──────────────────────────────────────────────────────┘
```

Figure 1.17

(B) (i)   Type f(g(x)) **ENTER**.
Explain what has happened.

(ii)   Repeat for $g(f(x))$, $h(g(x))$ and $g(h(x))$.

(iii)   Write down what you would expect to obtain for $f(h(x))$ and $h(f(x))$.

(C) (i)   Define further functions $s(x) = x^2$ and $r(x) = \sqrt{x}$.

(ii)   Type $s(g(x))$ and $r(g(x))$.

(iii)   Write down what you would expect for $s(r(x))$ and $r(s(x))$. Check your
predictions with the TI-92.

Composite Functions

When two functions are combined as in TI-92 Activity 1G a **composite
function** is formed. The rule is that the functions are applied in order from the right,
so for $fg(x)$, g is applied to x, and then the rule of f is applied to the result.

░░░░░░░░░░░░ **Example 1M** ░░░░░░░░░░░░

If $f(x) = x + 5$, $g(x) = x^2$ and $h(x) = x - 7$, find $fg(x)$, $gf(x)$ and $gh(x)$.

Solution

To find $fg(x)$ first note that $g(x)$ gives x^2. So

$$f(g(x)) = f(x^2) = x^2 + 5.$$

To find $gf(x)$ first note that $f(x)$ gives $x + 5$. So

$$g(f(x)) = g(x + 5) = (x + 5)^2.$$

To find $g(h(x))$ first note that $h(x)$ gives $x - 7$. So

$$g(h(x)) = g(x - 7) = (x - 7)^2.$$

░░

A simple diagrammatic view of a composite function is shown as a 'function machine'. Consider the composite function $fg(x)$ where f and g are defined by $f(x) = x + 5$ and $g(x) = x^2$. To evaluate the compositive $fg(x)$ we first carry out the function g on x and then carry out the rule for f on the result.

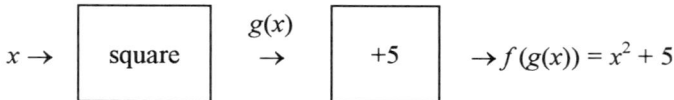

$$x \rightarrow \boxed{\text{square}} \xrightarrow{g(x)} \boxed{+5} \rightarrow f(g(x)) = x^2 + 5$$

To carry out the composite function $gf(x)$ we first carry out f on x and then apply g to the result

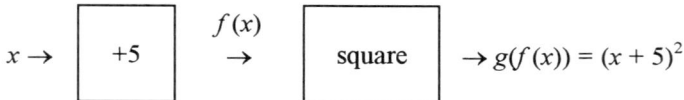

$$x \rightarrow \boxed{+5} \xrightarrow{f(x)} \boxed{\text{square}} \rightarrow g(f(x)) = (x + 5)^2$$

Exercise 1H

1.   Each of the following functions has domain $-3 \le x \le 5$. For each one find the range of the function and whether the mapping is one-to-one or many-to-one.

    (a) $f(x) = x^3$ (b) $f(x) = (x-3)^2$

    (c) $g(x) = 5 - x$ (d) $g(x) = 4/x$

    (e) $f(x) = x^4 + 1$ (f) $g(x) = |x - 2|$

2.   If $f(x) = x^2$, $g(x) = \dfrac{x}{2}$ and $h(x) = x - 6$, find

    (a) $f(0)$ (d) $f(2)$ (g) $f(y)$

    (b) $g(2)$ (e) $g(10)$ (h) $g(u)$

    (c) $h(9)$ (f) $h(-1)$ (i) $h(y^2)$

3.   For f, g and h as defined in problem 2, find

    (a) $fg(x)$ (d) $hg(x)$

    (b) $gh(x)$ (e) $fgh(x)$

    (c) $gf(x)$ (f) $ghf(x)$.

Graphs of Functions

Figure 1.18 shows the graphs of four functions $f(x) = x^2 - 1$, $g(x) = 1 - x^2$, $h(x) = x^3$ and $k(x) = x^3 - 3x^2 + 2$.

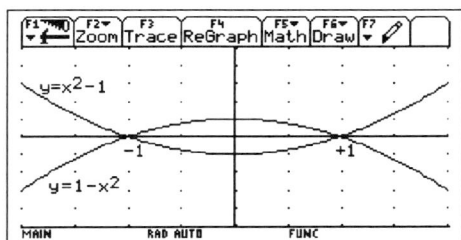

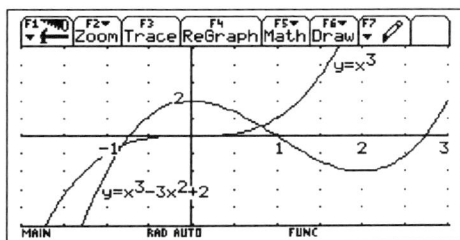

Figure 1.18

If you were asked to describe the graphs giving their essential features you would probably include the following points:

1.   The graphs of $f(x)$ and $g(x)$ each cut the x-axis at the points $x = -1$ and $x = 1$. The graph of $h(x)$ cuts the x-axis at $x = 0$. The graph of $k(x)$ cuts the x-axis in three points $x = -0.732$, $x = 1$ and $x = 2.732$. Such points, where the graph of a function $f(x)$ cuts the x-axis, are called the **roots** of the function. At a root $f(x) = 0$.

2.   The graph of $f(x) = x^2 - 1$ has **a minimum** at the point $(0,-1)$. It continuously decreases up to the point $(0,-1)$ and then increases as x continues to increase. The graph of $k(x) = x^3 - 3x^2 + 2$ has a similar feature at the point $(2,-2)$ but we cannot call it the minimum because for example at the point $(-2,-18)$ $f(x)$ is less than -2, i.e. 'more of a minimum value'. We call such a point **a local minimum**. What this means is that in the locality of $(2,-2)$ there is a minimum value for the function.

3.   The graph of $f(x) = 1 - x^2$ has **a maximum** at the point $(0,1)$. It continuously increases up to the point $(0,1)$ and then decreases as x continues to increase. The similar feature of the graph of $k(x)$ at the point $(0,2)$ is called **a local maximum**. Local maximum and local minimum points are collectively called **turning points**.

4.   The graph of $h(x) = x^3$ has a special feature at $(0,0)$. As x increases from say $x = -3$ the graph appears to be turning to be forming a maximum; but as x decreases from say $x = 3$ the graph appears to be turning to be forming a minimum. If we draw the tangent to the graph at the origin then the graph crosses its tangent (see Figure 1.19). Such a point $(0,0)$ is called a **point of inflexion**. For $y = x^3$ we say that the graph changes from 'concave down' to 'concave up'. We investigate points of inflexion more fully in Chapter 6.

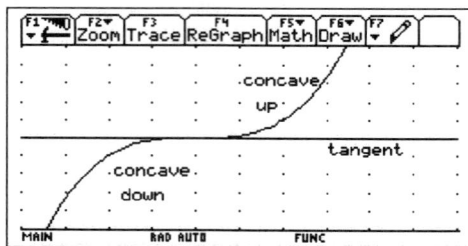

Figure 1.19   The graph crosses its own tangent at a point of inflexion

     To describe the graph of a function we note its features as the direction of x increases i.e. we scan the graph moving from left to right.

TI-92 ACTIVITY 1H

The aim of this Activity is to explore the features of graphs of polynomial functions. Clear the HOME screen, Y = Editor and GRAPH screen.

(A)   Set up and graph the function $y1 = 2x^4 + x^3 - 6x^2 + 1$ using WINDOW

$$xmin = -3, xmax = 3, xscl = 0.5,$$
$$ymin = -6, ymax = 6, yscl = 1.0$$

Use the cursor (**F3 Trace**) to describe the features of the graph, working to two decimal places.
What is the degree of the polynomial?
How many roots does the polynomial have?
How many local maximum and local minimum points does the graph of the polynomial have?

Table 1.8 summarises the description of the graph of the polynomial of degree 4, $y1(x) = 2x^4 + x^3 - 6x^2 + 1$.

Table 1.8

roots	turning points	
	type	coordinates
$x = -1.96, x = -0.41$	(local) minimum	$(-1.43, -5.83)$
$x = 0.44, x = 1.49$	local maximum	$(0, 1)$
	local minimum	$(1.05, -2.03)$

Note that in this case the point $(-1.43, -5.83)$ is actually the minimum value for the function. However we would usually still call it a local minimum.

(B)   Repeat Activity (A) for the following polynomials. (You may need to adjust the WINDOW parameters.)

(a) $x^4 - 2x^3 - x^2 + 2x$ (b) $2 - 5x + 5x^3 - 2x^4$

(c) $x^4 - 4x^2 - 1$ (d) $3x^2 + 2x^2 - 7x + 2$

(e) $x^3 - 2x$ (f) $1 + 2x - x^3$

(g) $5x^5 - 24x^4 + 20x^3 + 30x^2 - 25x - 6$

(h) $x^6 - 7.5x^4 + 12x^2 - 1$

From your results deduce a rule for the maximum number of roots and the maximum number of turning points for a polynomial of degree n.

(C)   Clear the Y = Editor and the GRAPH screen.
        Set up and graph the functions $y1(x) = x^2 - 2x + 2$, $y2(x) = x^2 - 2x + 1$ and $y3(x) = x^2 - 2x$.
        Describe the properties of each function.

        Figure 1.20 shows the GRAPH screen for this activity.

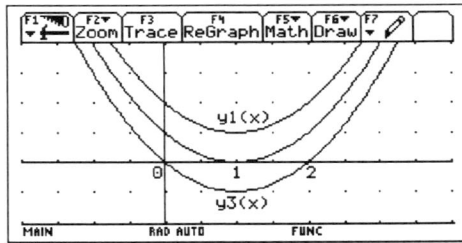

Figure 1.20   GRAPH screen for $y1(x) = x^2 - 2x + 2$, $y2(x) = x^2 - 2x + 1$ and $y3(x) = x^2 - 2x$

We see that for $y3(x)$ there are two roots $x = 0$ and $x = 2$.
For $y1(x)$ there are no points for which the graph cuts the x-axis. We will see in Chapter 10, that there are still two roots in terms of complex numbers.
For $y2(x)$ the graph 'touches the x-axis' at $x = 1$. We say that $x = 1$ is a **repeated root**.

(D)   Clear the Y = Editor and the GRAPH screen.
        Set up and graph the functions $y1(x) = x^3 - 3x^2 + 2x$, $y2(x) = x^3 - 2x^2 + x$, $y3(x) = x^3 - x^2 + x$ using the WINDOW

        xmin = -1, xmax = 2, xscl = 0.5,
        ymin = -1, ymax = 1, yscl = 0.5

Describe the properties of each function.
Sketch the possible shapes of graphs of cubic polynomials.

(E)   Consider the quartic polynomial $y = ax^4 + bx^3 + cx^2 + dx + e$.
Investigate the graphs of a quartic polynomial by choosing different values of the coefficients.
Sketch the possible shapes of graphs of quartic polynomials.

Summary

The graphs of the quadratic and cubic polynomials contain the essential features of the graph of other polynomials. Their properties are summarised in the following TI-92 screen dumps.

quadratic polynomial $f(x) = ax^2 + bx + c$

$a > 0$ concave up $a < 0$ concave down

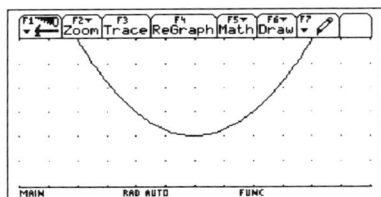

one (local) minimum   one (local) maximum

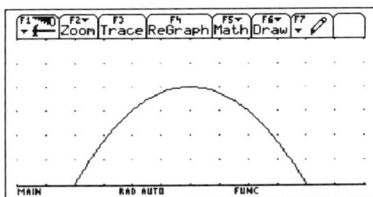

cubic polynomial $f(x) = ax^3 + bx^2 + cx + d$

$a > 0$ graph changes from concave down to concave up

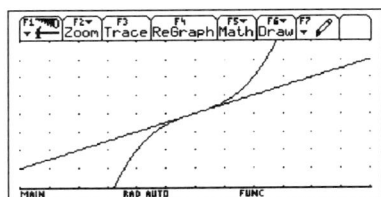

one point of inflexion   one local maximum
                                                one local minimum
                                                (one point of inflexion between)

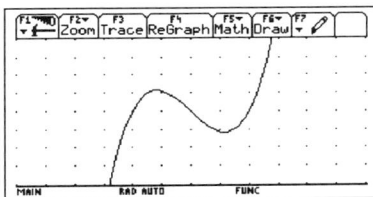

$a < 0$ graph changes from concave up to concave down

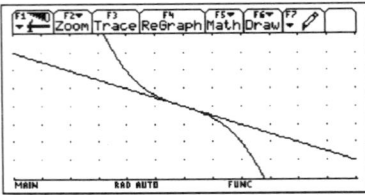

one point of inflexion   one local minimum
                                          one local maximum
                                          (one point of inflexion between)

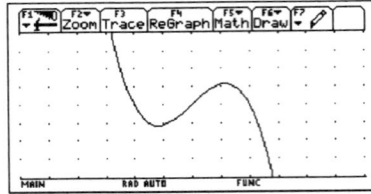

Inverse Functions

The **inverse of a function** will reverse the effect of the original function. If $f(x)$ is a function its inverse is written as $f^{-1}(x)$.

For example if $f(x) = x + 5$ then $f^{-1}(x) = x - 5$. Notice what happens to a value of x under f and then apply f^{-1} to the answer $f(x)$. For $x = 2$, $f(2) = 7$ and then $f^{-1}(7) = 2$. So $f^{-1}(f(2)) = 2$.

As another example, let $g(x) = \dfrac{x}{2}$ then $g^{-1}(x) = 2x$. If $x = 4$, then $g(4) = 2$ and $g^{-1}(2) = 4$ so that $g^{-1}(g(2)) = 2$.

Often functions are not this simple. Consider,

$$h(x) = \frac{(x-2)^2}{5}$$

$h(x)$ could be represented by a flow chart as below.

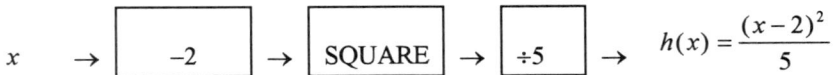

$x \quad \rightarrow \quad \boxed{-2} \quad \rightarrow \quad \boxed{\text{SQUARE}} \quad \rightarrow \quad \boxed{\div 5} \quad \rightarrow \quad h(x) = \dfrac{(x-2)^2}{5}$

The inverse can be obtained by reversing the flow chart and using the inverse of the individual functions as shown below.

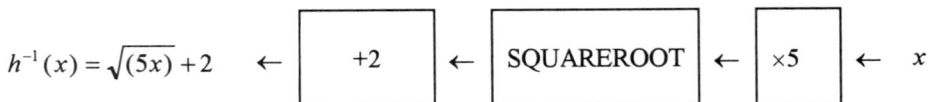

$h^{-1}(x) = \sqrt{(5x)} + 2 \quad \leftarrow \quad \boxed{+2} \quad \leftarrow \quad \boxed{\text{SQUAREROOT}} \quad \leftarrow \quad \boxed{\times 5} \quad \leftarrow \quad x$

Forming the composite of h and h^{-1} gives

$$h^{-1}(h(x)) = h^{-1}\left(\frac{(x-2)^2}{5}\right) = \sqrt{\frac{5(x-2)^2}{5}} + 2$$

$$= x$$

In general if the inverse of a function $f(x)$ is denoted by $f^{-1}(x)$ then

$$f^{-1}(f(x)) = x \text{ and } f\left(f^{-1}(x)\right) = x.$$

TI-92 ACTIVITY 1I

The aim of this Activity is to explore the graphs of a function and its inverse.

(A)   (i)   Using the Y = Editor, enter and graph x^2 and $\sqrt{x}$, a function and its inverse. Choose **F2 Zoom** and select 5:ZoomSqr. Graph the line $y = x$. What is the relationship between the graph of the function and its inverse?

       (ii)   Repeat for $5x$ and $x/5$.

       (iii) Repeat for $x + 1$ and $x - 1$.

(B)   (i)   Explore other functions and their inverses.
              The TI-92 has a useful built-in command for drawing the inverse function. Enter $y1 = x^2$ and graph. From the graph screen press **F6** and select 3:DrawInv (this sends you to the HOME screen). Type the function x^2 and press **ENTER** (this returns to the GRAPH screen and draws the inverse of x^2 i.e. $\sqrt{x}$). Figure 1.21 shows the screen dump for this activity.

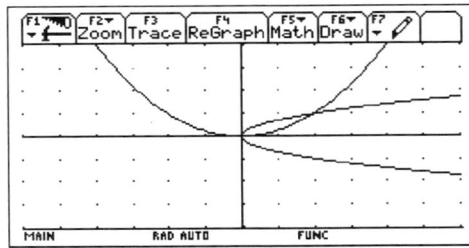

Figure 1.21

(C)   (i)   To obtain the inverse of functions such as $f(x) = 2x + 1$ type solve $(y = 2x + 1, x)$. This will produce the inverse function $f^{-1}(y)$ i.e. using y instead of x. Figure 1.22 shows this method for finding the inverse of the functions $f(x) = 2x + 1$ and $f(x) = \dfrac{x+3}{x-1}$.

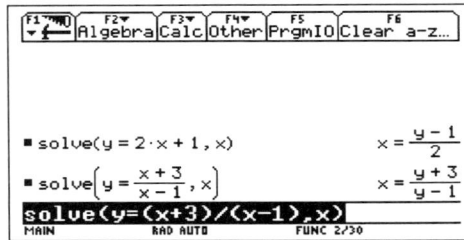

Figure 1.22

   (ii)   Repeat the process above for each of the following functions.

(a) $f(x) = 6x + 2$ (d) $f(x) = \dfrac{1}{4-x}$

(b) $f(x) = \dfrac{1}{x+2}$ (e) $f(x) = \dfrac{1}{x-3}$

(c) $f(x) = \dfrac{x+2}{x-1}$ (f) $f(x) = \sqrt{(5-x)}$

(iii) Now find and graph the inverse functions for each of the following functions.

(a) $f(x) = x^2 + 8$ (c) $f(x) = 4(x-7)^2$

(b) $f(x) = x^2 - 8$ (d) $f(x) = \dfrac{x^2}{1-x^2}$

Explain why two expressions are given when you use **F2** 1:solve(command.

Finding Inverse Functions

It is possible to find some inverse functions by using flow charts, but this method will not always work. The following Examples show an algebraic method of approach.

Example 1N

Find the inverse of the function

$$f(x) = \frac{x}{x-2}$$

Solution

First write the function in the form

$$y = \frac{x}{x-2}$$

Now this equation must be rearranged in the form $x = \ldots$

First multiply both sides by $(x-2)$.

$$(x-2)y = \frac{x(x-2)}{(x-2)}$$
$$xy - 2y = x.$$

Now taking all the terms that contain x to the left hand side (LHS) and those that do not contain x to the right hand side (RHS) gives

$$xy - x = 2y$$
$$x(y-1) = 2y$$

Now dividing by $(y-1)$ gives

$$x = \frac{2y}{y-1}$$

So the inverse function is given by

$$f^{-1}(x) = \frac{2x}{x-1}$$

Example 1P

Find the inverse of the function

$$g(x) = \frac{x+3}{x-1}$$

Solution

First the function is written in the form

$$y = \frac{x+3}{x-1}$$

Then it must be rearranged to make x the subject, i.e. in the form $x = \ldots$

First multiply both sides by $(x-1)$ to give

$$y(x-1) = x + 3$$

or

$$xy - y = x + 3 .$$

Then take all the terms that contain x to the LHS and all those that do not contain x to the RHS giving

$$xy - x = y + 3$$

or $x(y-1) = y + 3$

Now dividing by $(y-1)$ gives

$$x = \frac{y+3}{y-1}$$

Now the inverse function can be written as

$$g^{-1}(x) = \frac{x+3}{x-1}$$

Exercise 1I

1. Find the inverse of each function below by drawing flow charts.

(a) $f(x) = 6x - 10$ (e) $f(x) = \sqrt{2x-4}$

(b) $f(x) = 4(x+2)$ (f) $f(x) = 4\sqrt{(x+1)} - 5$

(c) $f(x) = \sqrt{(2x+7)}$ (g) $f(x) = \sqrt{\left(\frac{x}{2}+1\right)} - 5$

(d) $f(x) = 8x^3 - 5$ (h) $f(x) = \frac{\sqrt{x^3+4}}{5}$

2. Find the inverse of each of the following functions.

(a) $f(x) = \frac{x-4}{x+2}$ (d) $f(x) = \frac{1}{x} + \frac{1}{2x}$

(b) $f(x) = \frac{x}{1+x}$ (e) $f(x) = \frac{x-6}{x+2}$

(c) $f(x) = \frac{x^3}{x^3+2}$ (f) $f(x) = \frac{x^3+3}{2x^3+1}$

3. The functions f, g and h are defined as $f(x) = x^3$, $g(x) = \frac{1}{x+6}$ and $h(x) = 2x - 5$. Find

(a) $f^{-1}(0)$ (d) $g^{-1}(8)$

(b) $f^{-1}g(2)$ (e) $g^{-1}(-2)$

(c) $h^{-1}(4)$ (f) $h^{-1}(0)$.

1.10 DISCONTINUOUS FUNCTIONS

Functions can be formed by fractions where the denominator can be zero for certain values. For example,

$$y = \frac{6}{x-1}$$

will have a zero denominator when $x = 1$, and then y will be infinite. The value of $x = 1$ is called a **discontinuity**. The line $x = 1$ is called a **vertical asymptote**. The TI-92 Activity 1J will help you to discover the properties of some of these discontinuous functions.

TI-92 ACTIVITY 1J

(A)   (i)   Graph $\frac{1}{x}$, selecting ranges $-4 \le x \le 4$ and $-4 \le y \le 4$.

At what value of x does the function have its discontinuity?
What happens to the function for large positive or negative x values?

       (ii)   Repeat (i) for each function below.

              (a) $\dfrac{1}{x+1}$ (b) $\dfrac{1}{x-1}$ (c) $\dfrac{1}{x+2}$

       (iii) For what values of x will the functions below have discontinuities?

              (a) $\dfrac{1}{x-4}$ (b) $\dfrac{1}{x+10}$ (c) $\dfrac{1}{x-3}$

              Check your answers by plotting with suitable scales.

(B)   (i)   Clear the HOME screen and the Y = Editor.

       (ii)   Graph

              $$\frac{1}{(x-1)(x+2)}$$

              for ranges $-4 \le x \le 4$ and $-4 \le y \le 4$. Where is this function discontinuous? What happens to the function for large positive or negative values of x? For what range of values is the function negative?

(iii) Repeat (ii) for each function below. You may need to select different ranges in WINDOW.

(a) $\dfrac{1}{(x+1)(x-3)}$ (c) $\dfrac{1}{(2x-5)(3x-2)}$

(b) $\dfrac{1}{(x-2)(x-4)}$ (d) $\dfrac{1}{x(x-1)}$

(iv) Predict where each function below will have discontinuities.

(a) $\dfrac{1}{(x+1)(x-7)}$ (c) $\dfrac{1}{x(x-4)}$

(b) $\dfrac{1}{(2x+3)(x+4)}$ (d) $\dfrac{1}{(2x-5)(x+2)}$

Check your answers by plotting each function. Clear the Y = Editor and GRAPH screen.

(C) (i) Graph

$$\frac{x+1}{x-2}$$

for $-5 \le x \le 5$ and $-5 \le y \le 5$. Where is there a discontinuity of this function? What happens to the function for large positive or negative values of x? (Graph the function $y = 1$).

It appears that the function gets closer and closer to 1 for large values of x. Table 1.9 illustrates what happens as x increases.

Table 1.9

x	$f(x) = \dfrac{x+1}{x-2}$
1	-2
10	1.375
100	1.03061
1000	1.00301

As x grows larger and larger, the value of the function approaches the value 1.0. This written as

$$f(x) \to 1.0 \quad \text{as} \quad x \to \infty$$

Complete Table 1.10 to show what happens as x approaches $-\infty$.

Table 1.10

x	$f(x) = \dfrac{x+1}{x-2}$
-1	
-10	
-100	
-1000	

What do you deduce? $f(x) \to ?$ as $x \to -\infty$.

The line $y = 1$ is called a **horizontal asymptote**. The graph and asymptotes are shown in Figure 1.23.

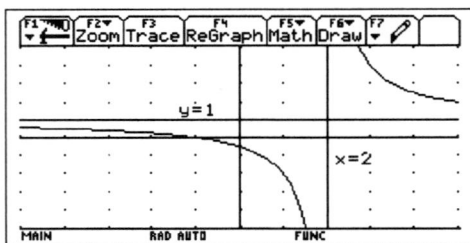

Figure 1.23

(ii)   Repeat (i) for the functions below.

(a) $\dfrac{2x+4}{x-1}$ (c) $\dfrac{(x-3)(x+4)}{(x+1)(2x-4)}$

(b) $\dfrac{3x-5}{6x+7}$ (d) $\dfrac{(x-1)(x+1)}{(2x-1)(2x+1)}$

Summary

TI-92 Activity 1J has introduced the idea of an asymptote and shown graphs of functions with discontinuities. To find the discontinuities we set the denominator to zero and solve for x.

To find any horizontal asymptotes we introduce the idea of the limit of the function as x gradually increases to $+\infty$ or gradually decreases to $-\infty$. For example, consider the function $f(x) = \dfrac{(x+1)^2}{(2x-3)(x+2)}$. Table 1.11 illustrates what happens as x increases.

Table 1.11

x	$f(x)$
1	−1.333
10	0.593
100	0.508
1000	0.501

As x grows larger and larger, the value of the function approaches the value 0.5. This is written as

$$f(x) \rightarrow 0.5 \quad \text{as} \quad x \rightarrow \infty.$$

Similarly Table 1.12 shows what happens as x approaches $-\infty$.

Table 1.12

x	$f(x)$
−1	0
−10	0.440
−100	0.493
−1000	0.499

We deduce

$$f(x) \rightarrow 0.5 \quad \text{as} \quad x \rightarrow -\infty \text{ and that } y = 0.5 \text{ is a horizontal asymptote.}$$

The vertical asymptotes are $x = 1.5$ and $x = -2$. Figure 1.24 shows a sketch of the function and the vertical asymptotes.

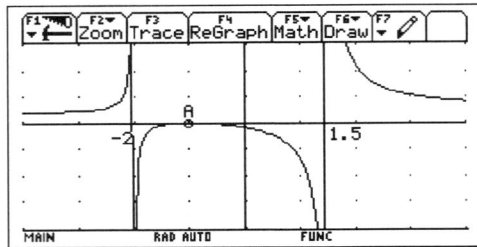

Figure 1.24 Graph of $f(x) = \dfrac{(x+1)^2}{(2x-3)(x+2)}$

In Chapter 6 you will learn how to sketch graphs of such functions by finding other properties of functions such as the coordinates of the turning point A.

Exercise 1J

Find any horizontal and vertical asymptotes of the following functions. Use the TI-92 to obtain graphs of the functions to confirm your answers.

(a) $y = \dfrac{3}{x-1}$

(b) $y = \dfrac{6}{3x+2}$

(c) $y = \dfrac{1}{x-1} + \dfrac{1}{x-2}$

(d) $y = \dfrac{1}{x} + \dfrac{1}{x+2}$

(e) $y = \dfrac{2x}{x+1}$

(f) $y = \dfrac{x}{x-4}$

(g) $y = \dfrac{3}{(x+2)(x-4)}$

(h) $y = \dfrac{x^2}{(x-5)(x+4)}$

1.11   INVESTIGATING LIMITS

At the end of the last section we explored the graph of the function $f(x) = \dfrac{(x+1)^2}{(2x-3)(x+2)}$. This is an example of a **rational function**. A rational function is a quotient of two functions in which both numerator and denominator are polynomials.

For the rational function $f(x)$ above, we have seen that there are discontinuities at $x = 1.5$ and $x = -2$. Furthermore, $f(x)$ tends towards 0.5 as x tends towards $+\infty$ and $-\infty$.

We introduce a special language and notation for this process. We say that $f(x)$ **tends to the limit 0.5**, as x tends to $\pm\infty$. This is written as

$$\lim_{x \to -\infty} f(x) = 0.5 \quad \text{and} \quad \lim_{x \to \infty} f(x) = 0.5$$

For the discontinuities the behaviour of the function is different. Consider again the graph in Figure 1.24 and the discontinuity at $x = 1.5$. On the left of $x = 1.5$ the value of $f(x)$ approaches $-\infty$ and on the right of $x = 1.5$ the value of $f(x)$ approaches $+\infty$. It is necessary to introduce a notation for this idea. As x approaches 1.5 from values smaller than 1.5 e.g. 1.49; 1.499; 1.4999 etc. we say that 'x **approaches 1.5 from below**' and write

$$x \to 1.5^-$$

As x approaches 1.5 from values larger than 1.5 e.g. 1.51; 1.501; 1.5001; etc. we say that 'x **approaches 1. 5 from above**' and write

$$x \to 1.5^+$$

Then

$$\lim_{x \to 1.5^+} f(x) = +\infty \quad \text{and} \quad \lim_{x \to 1.5^-} f(x) = -\infty$$

Common factors

Consider the function $g(x) = \dfrac{x^2 + x - 2}{x - 1}$. At first sight there appears to be a discontinuity at $x = 1$. However, for $x = 1$, $g(1) = \dfrac{0}{0}$ which is undefined.

To investigate what is happening here, Table 1.13 shows the limiting process as x approaches 1 'from below' and 'from above'.

Table 1.13

limit from above		limit from below	
x	$g(x)$	x	$g(x)$
1.1	3.1	0.9	2.9
1.01	3.01	0.99	2.99
1.001	3.001	0.999	2.999
1.0001	3.0001	0.9999	2.9999

The result of Table 1.12 suggest that

$$\lim_{x \to 2^-} g(x) = 3 \qquad \lim_{x \to 2^+} g(x) = 3$$

To understand what is happening here we can factorise the numerator to give $x^2 + x - 2 = (x + 2)(x - 1)$ so that

$$g(x) = \frac{(x+2)(x-1)}{(x-1)}$$

and dividing top and bottom by $(x - 1)$ (provided $x \ne 1$)

$$g(x) = x + 2 \qquad x \ne 1$$

The point at $x = 1$ with coordinates (1,3) is called a **removable discontinuity**. In fact the graph of $g(x)$ is the straight line $y = x + 2$.

░░░░░░░░░░░░░ **Example 1Q** ░░░░░░░░░░░░░

Find the vertical asymptotes of the graph of the rational function

$$f(x) = \frac{x^2 + 3x - 4}{x^2 - 1}$$

Solution

The denominator is zero when $x = -1$ and $x = 1$. For $x = -1$ the numerator has value -6 so that $x = -1$ is a vertical asymptote. For $x = 1$ the numerator also has a value of 0. So $(x - 1)$ is a factor of $x^2 + 3x - 4$. Factorising the numerator and denominator

$$y = \frac{x^2 + 3x - 4}{x^2 - 1} = \frac{(x+4)(x-1)}{(x+1)(x-1)}$$

and dividing top and bottom by $(x - 1)$ (provided $x \ne 1$)

$$y = \frac{x+4}{x+1} \qquad x \ne 1$$

The graph of this function shows that $x = -1$ is a vertical asymptote but $x = 1$ is not a vertical asymptote. At $x = 1$ there is a **removable discontinuity**.

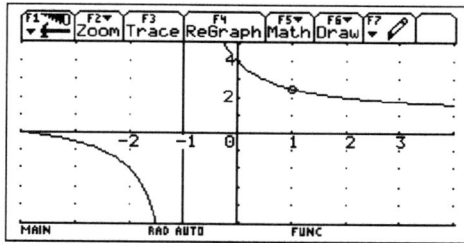

Figure 1.25   Graph of $y = \dfrac{x+4}{x+1}$

TI-92 ACTIVITY 1K

The aim of this Activity is to investigate limits using the TI-92.
Clear the HOME screen and the Y = Editor.

(A)   Consider in the function $f(x) = \dfrac{x^2-1}{x-1}$.   Use **F4**, select 1:Define and input
$f(x) = (x \wedge 2 - 1)/(x - 1)$.

    (i)   The limit command is in the F3 menu. Suppose we want to evaluate

$$\lim_{x \to 2^+} f(x)$$

        then press **F3** select 3:limit(and complete the line

           limit($f(x),x,2,1$) **ENTER**

        This evaluates the limit of $f(x)$ as x tends to 2 'from above'. To evaluate
$\lim\limits_{x \to 2^-} f(x)$ input

           limit($f(x),x,2,-1$) **ENTER**

        The -1 means x tends to 2 'from below'. Figure 1.26 shows the screen
dump for this activity. In each case the limit of $f(x)$ is 3.

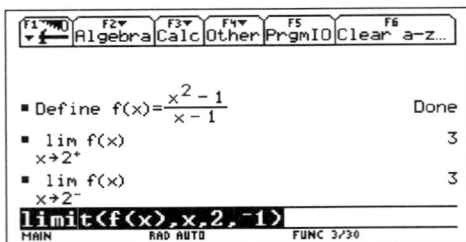

Figure 1.26   Finding the limit of $f(x)$

When using the TI-92 to evaluate the limit the 1 and –1 are optional if the limit from above equals the limit from below.

(ii)   Evaluate the following limits

$$\lim_{x\to1^-} f(x) \quad \text{and} \quad \lim_{x\to1^+} f(x)$$

Explain the significance of your answer. Is $x = 1$ a vertical asymptote?

Graph the function $\dfrac{x^2-1}{x-1}$. What is the equation of the line?

(B)   Repeat the procedure for each of the limits that follow. You need to evaluate each limit twice, once from above and once from below to see if there is any difference.

For clarity do not have more than one graph at a time in the GRAPH screen.

The most convenient WINDOW for each graph is given for you.

Which of them gives different answers on approaching from above and from below? Does the graph show you why?

    Expression   Suggested WINDOW

(a) $\displaystyle\lim_{x\to4}\frac{x^2-16}{x-4}$ $x$min = –6, $x$max = 6

                                $y$min = –6, $y$max = 6

(b) $\displaystyle\lim_{x\to3}\frac{x^3-8x-3}{x-3}$ $x$min = –6, $x$max = 6

                                $y$min = –20, $y$max = 20

(c) $\lim\limits_{x \to 3} \dfrac{1}{x-3}$ $x\text{min} = -10,\ x\text{max} = 10$

                                               $y\text{min} = -10,\ y\text{max} = 10$

(d) $\lim\limits_{x \to 0} \dfrac{4^x - 1}{x}$ $x\text{min} = -20,\ x\text{max} = 4$

                                               $y\text{min} = -1\ y\text{max} = 10$

(e) $\lim\limits_{x \to 0} x^x$ $x\text{min} = -4,\ x\text{max} = 4$

                                               $y\text{min} = -4,\ y\text{max} = 4$

Exercise 1K

1.   For the following functions determine whether there is a removable discontinuity at $x = 1$.

(a) $y = \dfrac{x^2 - 3x + 2}{x^2 - 1}$ (b) $y = \dfrac{x^2 + 3x + 2}{x^2 - 1}$

(c) $y = \dfrac{x^2 + 6x - 7}{x - 1}$ (d) $y = \dfrac{x^2 + 1}{x - 1}$

2.   Find the vertical and horizontal asymptotes of the function

$$y = \dfrac{4x}{\sqrt{x^2 - 4}}$$

What is the domain and range of this function?
Use the TI-92 to sketch the graph of this function to confirm your answers.

2

Exponential and logarithmic functions

2.1 INTRODUCTION

In Chapter 1 you have seen the properties of linear and polynomial functions and how these functions can be used to model physical situations. For example, $T = ke$ models the tension in an elastic spring for an extension e, and $s = \frac{1}{2}gt^2$ is the distance travelled by a ball when dropped vertically as a function of time t. In this chapter we explore other power laws.

As an example consider the motion of a simple pendulum consisting of a small heavy object attached to an inextensible string. Table 2.1 shows the period of the pendulum for different string lengths.

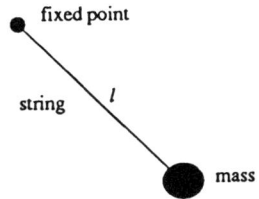

Figure 2.1   A simple pendulum

Table 2.1

Length ℓ (m)	0.5	0.6	0.7	0.8	0.9	1.0
Period T (s)	1.42	1.55	1.68	1.80	1.90	2.01

Figure 2.2 shows a graph of the period against the length. Clearly the graph is not linear and it does not have the general shape of a positive power of ℓ, examples of which are shown in Figure 2.3.

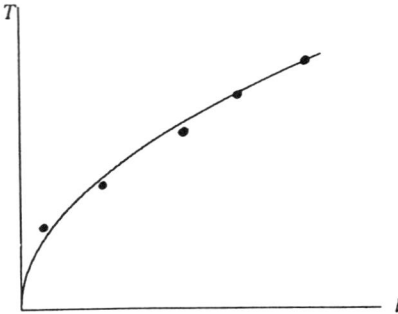

Figure 2.2

A graph of T against ℓ

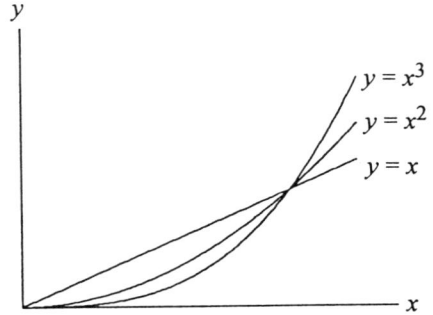

Figure 2.3

Graphs of $y = x$, $y = x^2$ and $y = x^3$

In fact Figure 2.2 is a graph of the form $\ell^{0.5}$ or $\sqrt{\ell}$. The actual model relating T and ℓ is

$$T = 2.01 \sqrt{\ell} \quad \text{or} \quad T = 2.01 \, \ell^{0.5}$$

A different type of power law occurs for the growth of money invested in a savings account. Suppose that you invest £3000 at 7% per annum. The growth of your investment is shown in Table 2.2.

Table 2.2

Time (years)	1	2	3	4	5
Amount (£)	3210	3434.7	3675.13	3932.39	4207.66

The model of this situation is

$$\text{amount} = 3000(1.07)^t$$

Notice that in this model the variable itself (i.e. t) is the power. This is often the situation in growth and decay processes.

The need to manipulate powers of ten occurs often in science and engineering because of the standard form of representing large and small numbers. For example, the mass of the earth is written as 5.98×10^{24} kg and the mass of an electron is 9.1×10^{-31} kg. This is certainly more convenient than writing 0.000 000 000 000 000 000 000 000 000 000 91 kg.

Each of these examples is of the form a^n which is called an **index law** or **exponential law**; a is called the **base** and n is called the **exponent** or **index** or **power**.

MINI-INVESTIGATION

1.   Using the TI-92 simplify the expressions below

 (a) $(x^2)(x^3)$ (b) $(x^6)(x^4)$ (c) $(x^m)(x^n)$

 What do you notice? Now try

 (d) $(x^4)(y^4)$ (e) $(x^7)(b^7)$ (f) $(z^2)(x^2)$

 What do you notice?

2.   Clear the HOME screen. Use the TI-92 to simplify

 (a) $(x^8) \div (x^5)$ (b) $(x^6) \div (x^8)$ (c) $(z^6) \div (z^4)$

 (d) $(x^3) \div (x^9)$ (e) $(x^6) \div (z^6)$ (f) $(z^4) \div (a^4)$

 What do you notice?

3.   Clear the HOME screen. Simplify

 (a) $(x^2)^4$ (b) $(x^3)^6$ (c) $(x^3)^4$ (d) $(x^n)^m$

 What do you notice?

4.   Clear the HOME screen. Simplify

 (a) 2^0 (b) x^0 (c) $x^{\frac{1}{2}}$

 (d) x^{-1} (e) x^{-2} (f) $x^{-\frac{1}{2}}$

Rules for manipulating indices

    product rule $a^m \times a^n = a^{m+n}$

    quotient rule $a^m \div a^n = \dfrac{a^m}{a^n} = a^{m-n}$

    power rule $(a^m)^n = a^{mn}$

Important conventions

$$\sqrt{a} = a^{\frac{1}{2}}$$

$$\sqrt[n]{a} = a^{\frac{1}{n}}$$

$$\frac{1}{x} = x^{-1}$$

$$\frac{1}{x^n} = x^{-n}$$

$$x^o = 1$$

Exercise 2A

1.   Simplify, using the three laws for indices and then check your answers with the TI-92.

(a) $x^3 \times x^2$ (b) $a^4 \times a^2 \times a^5$ (c) $(4a)^2 \times (2a)^3$

(d) $6x^5 \div 2x^2$ (e) $c^7 \div c^3$ (f) $(3^4)^2$

(g) $(a^2)^3$ (h) $(a^2 b^3)^4$ (i) 5^{-1}

(j) 2^{-4} (k) $a^{-1} \div a^{-3}$ (l) $\dfrac{4a^{-2} \times 2a^{-3}}{8a^{-4}}$

(m) $(mg)^2 \div (g/m)^{-1}$ (n) $a^x(a^{2x} - a^{-2x})$ (o) $\left(\dfrac{x^2 y}{a^3}\right)^{-2}$

2. Write each of the following in the form x^n

(a) $\sqrt{x}$ (b) $x^2\sqrt{x}$ (c) $\dfrac{x}{\sqrt{x}}$

(d) $\left(x^3\right)^2$ (e) $\left(\sqrt{x}\right)^4$ (f) $\dfrac{1}{x^5}$

(g) $\dfrac{x^2}{x^{-3}}$ (h) $\dfrac{x\sqrt{x}}{x^2}$ (i) $\dfrac{x^2\sqrt{x}}{x^3}$

3. Evaluate the following without using a calculator

(a) $9^{\frac{1}{2}}$ (b) $8^{\frac{1}{3}}$ (c) $16^{\frac{1}{4}}$

(d) $100^{\frac{3}{2}}$ (e) $\dfrac{1}{(1000)^{-\frac{1}{3}}}$ (f) $4^{-\frac{1}{2}}$

(g) $a^\circ \times b^\circ$ (h) $x^\circ + 3a^\circ + b^\circ$ (i) $\left(\dfrac{x^\circ}{a^\circ}\right)^{13}$

Now check your answers using the TI-92.

4. Write each of the following numbers as decimals to four significant figures.

(a) 3.2×10^5 (b) 1.473×10^{-4}

(c) 9.81×10^3 (d) 1.03×10^{-6}

(e) $(6.01 \times 10^4) \times (3.2 \times 10^6)$ (f) $(5.132 \times 10^9) \div (1.62 \times 10^3)$

(g) $(2.43 \times 10^5)^2$ (h) $(4.72 \times 10^6) + (1.96 \times 10^4)$

5.   Write the answers to each of the following in standard form $A \times 10^{n}$ where $|A| < 10$ and n is an integer.

(a)   The kinetic energy of a train of mass 300 000 kg moving with speed 50 ms^{-1}. Note that kinetic energy = $\frac{1}{2}mv^{2}$.

(b)   The wavelength associated with electrons is given by $\lambda = h / p$ where h is Planck's constant 6.63×10^{-34} Js and p is the momentum of the electrons. Calculate the wavelength for electrons with momentum 2.1×10^{-23} Ns.

(c)   Newton's law of gravitation gives the force on the moon due to the earth as

$$F = \frac{GMm}{r^{2}}$$

where G is the gravitational constant 6.67×10^{-11} Nm^{2} kg^{-2}, M is the mass of the earth 5.98×10^{24} kg and r is the distance of the moon from the centre of the earth $\mathbf{3}.84 \times 10^{8}$ m. Calculate F.

(d)   The kinetic energy of an electron of mass 9.1×10^{-31} kg moving with speed 2×10^{4} ms^{-1}.

(e)   The force on a rocket of mass 30000 kg accelerating at 21 ms^{-2}.

(Note that Newton's second law of motion gives force = mass × acceleration).

(f)   The number of seconds in one year (of 365 days).

2.2   THE EXPONENTIAL FUNCTION e^x

TI-92 ACTIVITY 2A

In this activity you will investigate exponential functions and their graphs.
Clear the HOME screen and the Y = Editor.

(A)   (i)   Define $y1 = 2^x$ and graph $y1$ with WINDOW

xmin = −1, xmax = 3, xscl = 0.5,
ymin = −1, ymax = 10, yscl = 1.0

Describe the properties of 2^x when (a) x is large and negative, (b) when
$x = 0$ and (c) when x is large and positive.

(ii)   Draw the graph of 3^x on the same grid as 2^x.
Repeat activity (ii) for the expressions 5^x and 10^x. What features do all
four graphs have in common?
How are they different? Clear the Y = Editor.

(B)   Repeat the same procedure for the following expressions

2^{-x}, 3^{-x}, 5^{-x} and 10^{-x}.

What features do these graphs have in common? How are they different? How
do they compare with the graphs drawn in (A)?

Summary

The exponential function $f(x) = a^x$ with $a > 1$ can be used to model quantities which
grow larger as x increases. The shape of the graph of a^x suggests that the graph
becomes more steep as x increases.

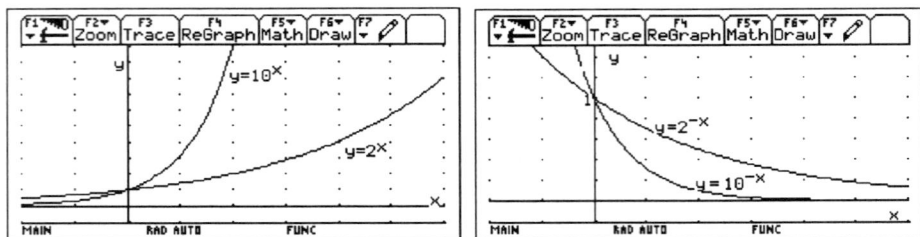

Figure 2.4   Graphs of $y = 2^x$, $y = 10^x$

Figure 2.5

Graphs of $y = (\frac{1}{2})^x = 2^{-x}$, $y = (\frac{1}{10})^x = 10^{-x}$

For $0 < a < 1$ the graph of $f(x) = a^x$ decreases as x increases so that a^x for $a < 1$ can be used to model quantities which decay. For example, $T(t) = 0.16^t$ could model the temperature of a bowl of hot soup as a function of time, t, as it cools down.

TI-92 ACTIVITY 2B

In this activity you will investigate the gradient of the graphs of exponential functions. The gradient of a graph at a point is the slope of the tangent to the graph at the point Select Approximate mode (using **MODE F2** Exact/Approx).

(A)   (i)   Set up $y1 = 2^x$ and graph 2^x.

      (ii)   The tangent to a graph at a given point can be drawn using **F5** A:Tangent in the GRAPH screen. Respond to 'Tangent at ?' with 0. Write down the slope of the tangent at $x = 0$.

      (iii)   Draw the tangent to 2^x at $x = 1$. Write down the slope of the tangent at $x = 1$.

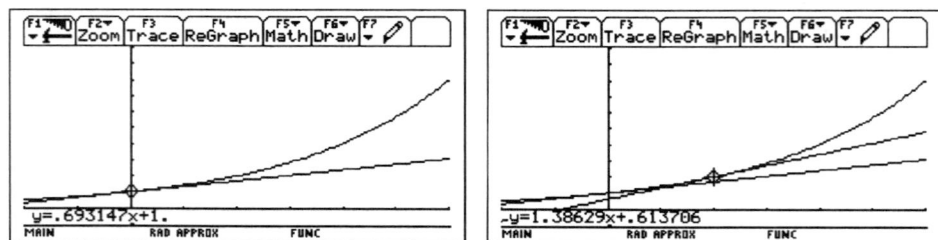

Figure 2.6   Tangent to graph of 2^x at $x = 0$ and $x = 1$

(iv)   Now complete the following table.

Table 2.3

x	$y = 2^x$	slope of tangent
0	1	0.693147
1	2	
2	4	
3	8	

(B)   (i)   Repeat activity (iv) for the function $y = 3^x$. Does your table confirm the statement in the summary that the graphs become more steep as x increases?

Note from your tables that for the function 2^x the slopes are less than the values of the function whereas for 3^x the slopes are greater than the function values.

(C)   (i)   The next function is rather special.
Clear the Y = Editor and GRAPH screen. Define $y1 = e^x$. This is a special function in mathematics called **the exponential function**. Graph e^x. Compare this graph with the graphs of 2^x and 3^x. What can you deduce about the properties of e^x?

(ii)   Use the Tangent Utility to draw the tangent and find the slope at each point to complete the following table.

Table 2.4

x	$y = e^x$	slope of tangent
0		1.0
1		
2		
3		

Comment on the results in your table. Find a rule for the slope of the tangent in terms of y.

(D)   Repeat activity (C) for the special function e^{-x}.

Summary

TI-92 Activity 2B introduces a very important function in mathematics denoted by e^x. It is called **the exponential function**. The number e is an irrational number (like for example π) and to 12 decimal places is given by

$$\boxed{e = 2.718281828459}$$

The important property of the exponential function that makes it one of the most useful models in mathematics was shown in activity (C). This property is associated with the slope of the tangent to the graph of a function $f(x)$ called **the rate of change of $f(x)$ with respect to x**.

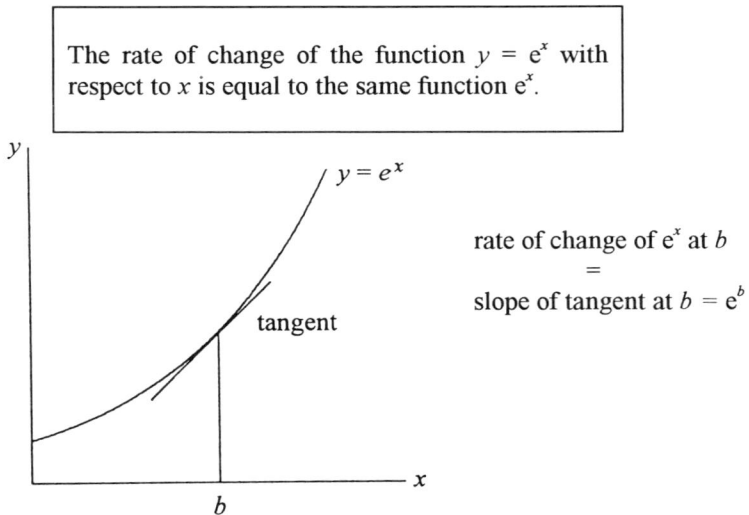

> The rate of change of the function $y = e^x$ with respect to x is equal to the same function e^x.

rate of change of e^x at b
=
slope of tangent at $b = e^b$

Figure 2.7

So what this means is that the rate of change of e^x with respect to x at $x = 2$ is actually equal to e^2. Compare this with the functions 2^x and 3^x say. The rate of change of 2^x is always less than the function value whereas the rate of change of 3^x is always greater than the function value.

In theory we can model any growth or decay function by a power law a^x for some number a. This turns out to be very inconvenient and it is more usual to use the base e.

Suppose that the model for a physical quantity is

$$y = Aa^x$$

This can be rewritten as

$$y = Ae^{kx}$$

where $a = e^k$ since $a^x = (e^k)^x = e^{kx}$.

The general shape of the graph of the functions $y = Ae^{kx}$ for $k > 0$ and $k < 0$ are shown in Figure 2.8.

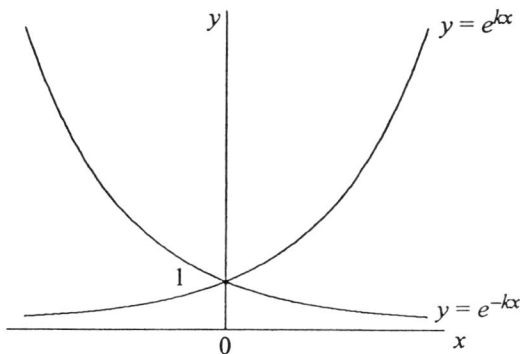

Figure 2.8   Graphs of e^{kx} for $k > 0$ and $k < 0$

Exercise 2B

1. Scientific calculators contain the exponential function as a special key. It is usually labelled e^x. Use your calculator to find the following as decimals to four significant figures.

   (a) e^0 (b) e^2 (c) e^{-3} (d) $e^{1.6}$ (e) $e^{0.21}$

   (f) $e^{0.6}$ (g) e^1 (h) $\sqrt{e}$ (i) e^4 (j) $e^{-0.2}$

2. Use the TI-92 to draw graphs of the following functions.

   (a) e^{2x} (b) $5e^{3x}$ (c) e^{-x} (d) $4e^{-2x}$

3. Use the Tangent Utility in the TI-92 to find the rate of change of the following as decimals to four significant figures.

   (a) 2^t at $t = 1.5$ (b) e^t at $t = 2.1$

   (c) e^{3x} at $x = 2$ (d) 4^{-x} at $x = 3$

4.   A function of importance in statistics is one of the form $f(x) = e^{-x^2}$. Use the
      TI-92 to sketch a graph of this function.

5.   Consider the expression $y = e^{3x}$. Use the Tangent Utility of the TI-92 to draw
      the tangent and find the slope at each point to complete the following table.

                                  Table 2.5

x	$y = e^{3x}$	slope of tangent
0		
1		
2		
3		

Find a rule for the rate of change of y in terms of y.

Investigate the expression $y = e^{ax}$ for different values of a. Deduce a rule for
the rate of change of e^{ax}.

TI-92 ACTIVITY 2C

In this activity you will use the TI-92 to investigate another important property of the
exponential function. Clear the HOME screen.

(A)   Press **F4**, select 1:Define and **ENTER**.
       Enter the expression $f(n) = (1 + 1/n) \wedge n$.
       Evaluate $f(1)$ and $f(1.5)$. You should have the screen shown in Figure 2.9.

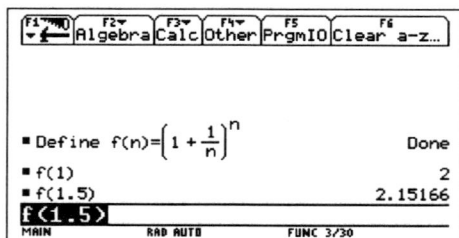

Figure 2.9

Complete Table 2.6 and show that as n increases, $f(n)$ tends towards e^1.

Table 2.6

n	$\left(1+\dfrac{1}{n}\right)^n$
1	2
1.5	2.1516574
2	
3	
4	
5	
10	
100	
1000	
10000	

(B)   Now investigate each of the following expressions as n increases.

$$\left(1+\frac{2}{n}\right)^n, \quad \left(1+\frac{3}{n}\right)^n, \quad \left(1+\frac{4}{n}\right)^n, \quad \left(1+\frac{5}{n}\right)^n$$

Find a rule between the limit of each expression and the exponential function e^x.

(C)   Imagine that you have £1000 to invest for one year at a nominal annual rate of interest of 8%.

(a)   How much would your investment be worth after one year if interest is compounded (i) annually, (ii) quarterly, (iii) monthly, (iv) weekly, (v) daily?

(b)   What would be the formula for continuous compounding?

2.3   LOGARITHMIC FUNCTIONS

Consider the exponential function $y = e^x$. If we are given a value of x, 1.5 say, then we can calculate y

$$y = e^{1.5} = 4.48168$$

However, suppose that we are given $y = 4.48168$ how do we show that $x = 1.5$? We would need to solve the equation

$$e^x = 4.48168$$

Figure 2.10 shows a diagram of the mathematical problem.

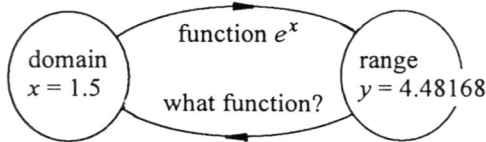

Figure 2.10

We are looking for the inverse function to e^x. For exponential functions the inverse functions are called **logarithmic functions**.

> ### Definition of a Logarithmic Function
>
> $$y = \log_a x \qquad a > 0, \, a \neq 1$$
>
> is the inverse of the function $x = a^y$.

Although the index a could take any value, there are two bases which are most commonly used in mathematics,

   $a = 10$ $\log_{10} x$ are called **common logarithms**

   $a = e$ $\log_e x$ are called **natural logarithms** and are written as $\ln(x)$.

Hence to solve the equation $e^x = 4.48168$ we use the natural logarithm to give $x = \ln(4.48168) = 1.5$.

TI-92 ACTIVITY 2D

In this activity you will use the TI-92 to investigate logarithmic functions and their graphs. Clear the HOME screen and Y = Editor. Set MODE to Auto.

(A) (i) Assign $y1 = \ln(x)$ and GRAPH. Describe the properties of $\ln(x)$

(ii) Assign $y2 = \log(x)$ and GRAPH. Describe the properties of $\log(x)$.

Compare the two graphs. What features are the same and what features are different? For what value of x do the graphs cut the x-axis? Use the definition of logarithmic functions to explain your answer.

(B) The TI-92 has two built in logarithmic functions $\ln(x)$ and $\log(x)$. We know that $y = \log_a x$ and $x = a^y$. On the TI-92, $a = e$ for $\ln(x)$ and $a = 10$ for $\log(x)$.

What about logarithmic functions to the base 2, 3, 4, ... for example $y = \log_2(x)$, $x = 2^y$.

In HOME screen enter Solve $(x = 2^y, y)$, in effect you are finding the inverse of $x = 2^y$, which we know is the definition of a logarithmic function, in terms of $\ln(x)$ and $\ln(2)$ (commands that the TI-92 can calculate).

You should obtain $y = \dfrac{\ln(x)}{\ln(2)}$. Graph this function. This is the same function as $y = \log_2(x)$. Repeat for (a) $y = \log_3(x)$, (b) $y = \log_4(x)$, (c) $y = \log_5(x)$ using the same method.

(C) (i) Clear the HOME screen and Y = Editor. Graph e^x and $\ln(x)$. Compare the graphs of e^x and $\ln(x)$. Graph $y = x$ (so that the line $y = x$ is 45° to the x-axis choose F2 5:ZoomSqr.) Note that the graph of $y = \ln(x)$ is a reflection of the graph of $y = e^x$ in the line $y = x$.

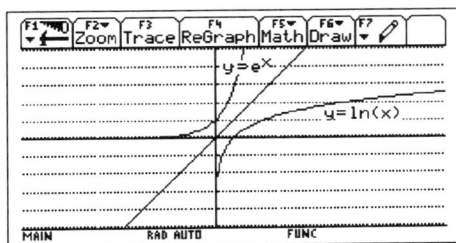

Figure 2.11

(ii)   Repeat activity (i) for 10^x and $\log(x)$. Do you see the same reflection property?

(D)   Clear the Y = Editor and delete the graphs. Assign $y1 = \dfrac{\ln(x)}{\ln(2)}$ $(= \log_2(x))$ and GRAPH. Graph the line $y = 1$.

Use the Trace and Zoom to move the cursor to find the value of x where the graphs cross. Repeat for $y = \log_3(x)$ and $y = \log_5(x)$. Deduce a general result for the solution of $\log_a(x) = 1$.

Summary

1.   The graph of each logarithmic function has the general shape shown in Figure 2.12.

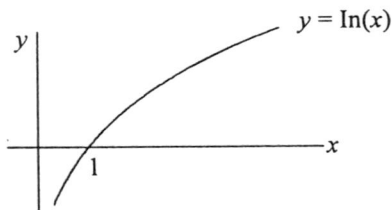

Figure 2.12   The graph of $y = \ln(x)$

2.   For any base $\log_a(1) = 0$ and $\log_a(a) = 1$.

Exercise 2C

1.   Common and natural logarithms are available on your calculator. They are usually labelled log and ln respectively.

Use your calculator to find the values of the following logarithms.

(a) $\log(2)$ (b) $\log(10)$ (c) $\log(4.2)$ (d) $\log(0.7)$

(e) $\ln(2)$ (f) $\ln(4.2)$ (g) $\ln(0.7)$ (h) $\ln(1)$

(i) $\log(1000)$ (j) $\ln(e^2)$ (k) $\log(1)$ (l) $\ln(e)$

2.   Use the definition of logarithmic functions to show that $\log_a(a) = 1$ for any base a. Check the result using the TI-92 and several values for a.

3.   Solve the following equations.

   (a) $3.1 = 10^x$ (b) $0.2 = 10^x$

   (c) $1.7 = e^x$ (d) $5.1 = e^x$

   (e) $11.2 = e^t$ (f) $0.47 = e^t$

   (g) $\ln(t) = 1.7$ (h) $\ln(x) = 4.2$

   (i) $\log(3x) = 6$ (j) $\ln(5t) = 3.1$

4.   The charge on a capacitor as it is charged by a battery is given by

$$Q = Q_0(1 - e^{-50t})$$

   where Q_0 is a constant. Interpret the quantity Q_0. Calculate the time when
   $Q/Q_0 = 0.5$. Use the TI-92 to draw a graph of charge against time when
   $Q_0 = 10$. Explore what happens to Q as t becomes large?

5.   Radioactive decay of substances is modelled by the exponential function. For a
   sample of iron nuclide the fraction remaining as a function of time is modelled
   by

$$N = e^{-0.25t}$$

   Calculate the time when $N = 0.5$. This is called the *half-life* of the decay.

6.   Radioactive carbon-14 decays at a rate of $1.238 \times 10^{-4}\,\text{yr}^{-1}$ so that the fraction
   of carbon-14 in a sample is modelled by

$$M = e^{-1.238 \times 10^{-4}t}$$

   Calculate the half life of carbon-14. Calculate the times for the fraction of
   carbon-14 in the sample to reduce to 10% and 5%.

7.   The atmospheric pressure at a height h (in km) above sea level obeys the rule

$$p = p_0 e^{-0.15h}$$

where p_0 is the pressure at sea level, $p_0 = 100\ 000$ pascals.

(a)   Calculate the pressure at heights 1 km, 2 km, 5 km and 10 km.

(b)   Draw a graph of p against h.

(c)   At what height is the pressure equal to half the value at sea level?

(d)   At what height is the pressure equal to one-tenth of the value at sea level?

2.4   RULES FOR LOGARITHMS

The following two properties of logarithms have already been observed.

$\ln(1) = 0$	$\log_{10}(1) = 0$
$\ln(e) = 1$	$\log_{10}(10) = 1$

MINI INVESTIGATION

(A)   Set MODE to EXACT. Clear the HOME screen.

(i)   Using the TI-92 simplify the following expressions.

(a) $\ln(2) + \ln(3)$ (b) $\ln(5) + \ln(10)$

(c) $\ln(6) + \ln(2)$ (d) $\ln(2) + \ln(x)$

What do you observe?

(ii)   Simplify the following

(a) $\ln(4) - \ln(2)$ (b) $\ln(8) - \ln(2)$

(c) $\ln(1000) - \ln(10)$ (d) $\ln(x) - \ln(3)$

What do you observe?

(iii)   Simplify

      (a) $3\ln(4)$ (b) $4\ln(2)$

      (c) $2\ln(9)$ (d) $\frac{1}{2}\ln(64)$

      (e) $3\ln(1)$ (f) $\frac{1}{5}\ln(32)$

      What do you observe?

(B)   Use the TI-92 to expand the following expressions.   (Use **F2** and select 3:expand.)

      (a) $\ln(1)$ (b) $\ln(9)$ (c) $\ln(2.5)$

      (d) $\ln(\sqrt{2})$ (e) $\ln(12)$ (f) $\ln(24)$

      (g) $\ln(2x)$ (h) $\ln\left(\dfrac{x}{5}\right)$ (i) $\ln\left(\dfrac{3x}{2}\right)$

(C)   Do the results change if logarithms to a different base are used?

Summary

You will have observed that there are three important rules for manipulating logarithms. These rules are:

product rule	$\log_a(xy) = \log_a x + \log_a y$
quotient rule	$\log_a(x/y) = \log_a x - \log_a y$
power rule	$\log_a(x^r) = r\log_a x$

To prove these three rules we use the rules for manipulating indices.
      Let $p = \log_a(x)$ and $q = \log_a(v)$, then using the definition of logarithms $x = a^p$ and $y = a^q$.

Product rule   Multiplying x and y we have

$$xy = a^p a^q = a^{p+q}$$

Then

$$\log_a(xy) = p + q$$
$$= \log_a(x) + \log_a(y)$$

Quotient rule   Dividing x by y we have

$$\frac{x}{y} = \frac{a^p}{a^q} = a^{p-q}$$

Then

$$\log_a(x/y) = p - q$$
$$= \log_a(x) - \log_a(y).$$

Power rule $x^r = (a^p)^r = a^{pr}$

Then

$$\log_a x^r = pr = r \log_a(x)$$

Exercise 2D

1.   Expand the following expressions using the rules of logarithms.

(a) $\ln(a^3 b^2)$ (b) $\ln(0.1x^2)$

(c) $\ln(3.7t/a)$ (d) $\ln(3p(x+y)^2)$

(e) $\log(\sqrt{2}s/t)$ (f) $\log(0.2v^2)$

(g) $\log(10xy^3)$ (h) $\ln(\frac{1}{2}at^2)$

(i) $\log(3(a+b)/s)$

2.   Write each of the following as a single logarithm and simplify.

    (a) $\log(40) - \log(5)$ (b) $\log(10x) + \log(2x) - \log(x)$

    (c) $\ln(5a) + \ln(2b) - \ln(c)$ (d) $3\ln(a) + 5\ln(b)$

    (e) $2\ln(x) - 4\ln(y)$ (f) $\log(x) - \log(y) + 1.5\log(a)$

3.   Solve the following equations.

    (a) $\ln(2x) = 5.6$

    (b) $2\ln(3x) - 3\ln(x) = 1.5$

    (c) $3\log(x) + 2\log(3x) = 3.1$

2.5   MODELLING WITH POWER AND EXPONENTIAL LAWS

Scientific laws usually arise in one of two ways. Either existing scientific theories are used to develop new theories which are tested by experiment or results of experiments are used directly to formulate empirical models. In both of these approaches the use of logarithms can play an important part.

In Chapter 1 we saw that if data for two variables fits a straight line then finding its equation and hence the relation between the variables is $y = mx + c$ where m is the slope of the line and c is the intercept. But experimental data does not always give straight lines. However, in situations where the model is a power law $y = ax^b$ or an exponential law $y = ae^{bx}$ the use of logarithms will transform the graphs to straight lines.

Power Law. Suppose that the relationship between two variables is thought to be a power law of the form $y = ax^b$. Then a graph of y against x will be of the form shown in Figure 2.13. An important clue to a power law is that the graph passes through the origin.

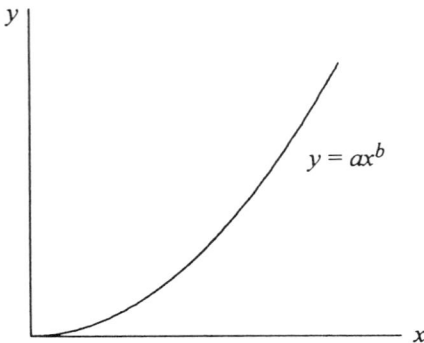

Figure 2.13

Typical power law graph $y = ax^b$

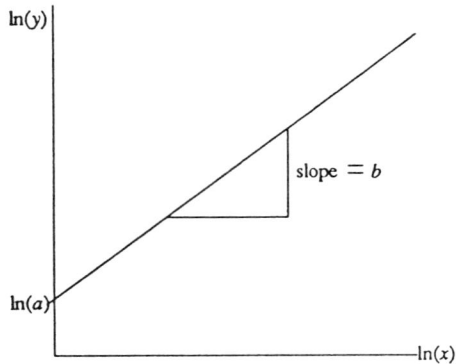

Figure 2.14

For a power law $y = ax^b$ a graph of $\ln(y)$
against $\ln(x)$ will be linear

If we take (natural) logarithms of each side of the equation we get

$$\ln(y) = \ln(ax^b) = \ln(a) + \ln(x^b)$$
$$= \ln(a) + b \ln(x)$$

Thus a graph of $\ln(y)$ against $\ln(x)$ will be a straight line (see Figure 2.14).

The slope of the straight line graph is the index b and the intercept $\ln(a)$ can be used to find a.

Exponential Law. Suppose that the relationship between the two variables x and y is thought to be exponential of the form $y = ae^{bx}$. Then a graph of y against x will be of the form shown in Figure 2.15.

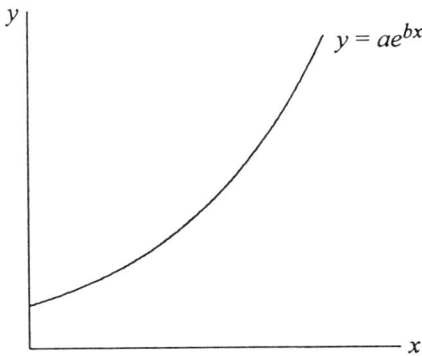

Figure 2.15
Typical exponential law graph $y = ae^{bx}$

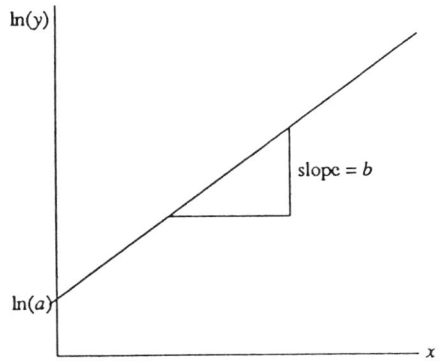

Figure 2.16
For an exponential law $y = ae^{bx}$
a graph of $\ln(y)$ against x will be linear

If we take (natural) logarithms of each side we get

$$\ln(y) = \ln(ae^{bx}) = \ln(a) + \ln(e^{bx})$$
$$= \ln(a) + bx.$$

Thus a graph of $\ln(y)$ against x will be a straight line. The slope and intercept of the straight line graph can be used to find a and b.

The following example shows how we use the TI-92 to find relationships between variables.

Example 2A

Table 2.7 shows three sets of experimental data, one pair satisfies a power law, one pair satisfies an exponential law and the other pair does not satisfy either of these laws. Use appropriate graphs to find the relationships.

Table 2.7

r	0.1	0.4	0.7	1.0	1.3	1.6
w	2.010	0.606	0.182	0.055	0.017	0.005

u	0.1	0.3	0.6	0.9	1.2	1.5
v	1.00	0.96	0.83	0.62	0.36	0.07

t	0.1	0.5	0.9	1.3	1.7	2.1
s	0.084	1.293	3.511	6.561	10.35	14.83

Solution

The first step is to enter the data into the TI-92 using the Data/Matrix Editor. In the screen dump in Figure 2.17 we have used c1 for r, c2 for w, c3 for u and so on.

Figure 2.17   The Data/Matrix Editor

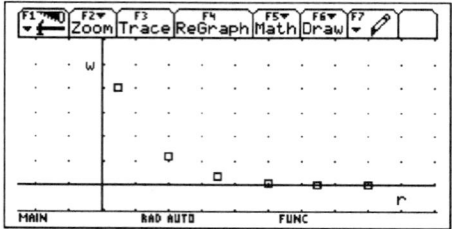

Figure 2.18   Graph of w against r

The graphs of each data set are shown in Figures 2.18, 2.19 and 2.20.

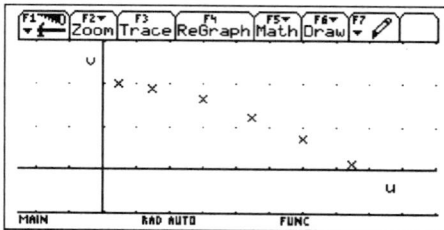

Figure 2.19   Graph of v against u

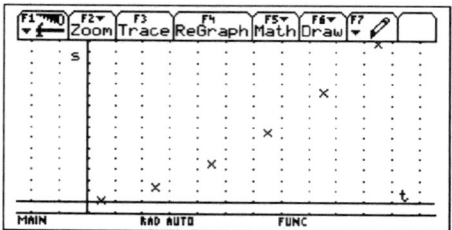

Figure 2.20   Graph of s against t

The graph in Figure 2.20 suggests that s and t are related by a power law $s = at^b$ and the graph in Figure 2.18 suggests that r and w are related by an exponential law $w = ae^{br}$. The relation between u and v in Figure 2.19 is neither of these.

Consider the r, w data. Since we expect an exponential law $w = ae^{br}$ we use LinReg calculation to find a linear law between $\ln(w)$ and r. Set up column c7 as $\ln(c2)$ by highlighting c7, press **F3** and complete the line $c7 = \ln(c2)$ (see Figure 2.21). Now press **F5** and calculate the LinReg with c1 for x and c7 for y. Figure 2.22 shows the results of the calculation.

	c1	c2	c3	c4	c5	c6	c7
1	.1	2.01	.1	1	.1	.084	.698
2	.4	.606	.3	.96	.5	1.29	-.5
3	.7	.182	.6	.83	.9	3.51	-1.7
4	1.	.055	.9	.62	1.3	6.56	-2.9
5	1.3	.017	1.2	.36	1.7	10.4	-4.1
6	1.6	.005	1.5	.07	2.1	14.8	-5.3
7							

c7=ln(c2)

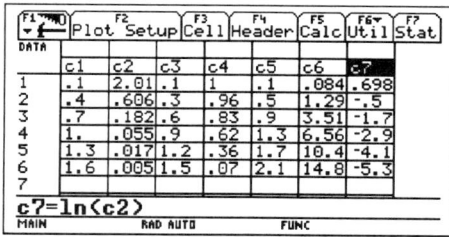

Figure 2.21   Setting up c7 = ln(w)

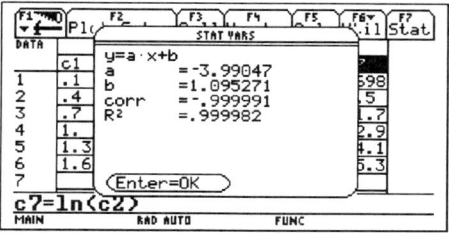

y=a·x+b
a   = -3.99047
b   = 1.095271
corr = -.999991
R²   = .999982

c7=ln(c2)

Figure 2.22   LinReg for ln(w) and r

We have

$$\ln(w) = 1.095271 - 3.99047r$$

Comparing with the general form on page 79

$$\ln(y) = \ln(a) + bx$$

we deduce that $b = -3.99$ and $\ln(a) = 1.095$ i.e. $a = e^{1.095} = 2.99$ giving the relationship

$$w = 2.99e^{-3.99r}$$

[It is likely that the law $w = 3e^{-4r}$ would be a very good model for this data.]

Consider the t, s data. Since we expect a power law $s = at^b$ we use LinReg calculation to find a linear law between $\ln(s)$ and $\ln(t)$. Figure 2.23 shows the screen for this activity. We have defined c8 (= ln(c5) = ln(t)) and c9 = ln(c6) (= ln(s)).

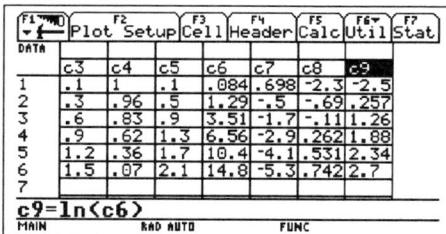

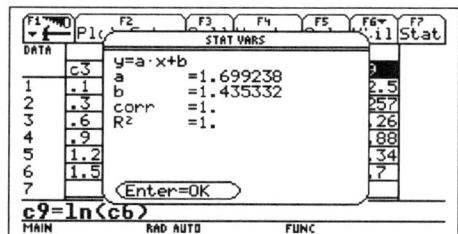

	c3	c4	c5	c6	c7	c8	c9
1	.1	1	.1	.084	.698	-2.3	-2.5
2	.3	.96	.5	1.29	-.5	-.69	.257
3	.6	.83	.9	3.51	-1.7	-.111	1.26
4	.9	.62	1.3	6.56	-2.9	.262	1.88
5	1.2	.36	1.7	10.4	-4.1	.531	2.34
6	1.5	.07	2.1	14.8	-5.3	.742	2.7
7							

c9=ln(c6)

y=a·x+b
a   = 1.699238
b   = 1.435332
corr = 1.
R²   = 1.

c9=ln(c6)

Figure 2.23   LinReg for ln(s) and ln(t)

We have

$$\ln(s) = 1.699238\ln(t) + 1.435332$$

Comparing this with the general form on page 78

$$\ln(s) = \ln(a) + b\ln(t)$$

We deduce that $b = 1.699$ and $\ln(a) = 1.435$ so that $a = e^{1.435} = 4.2$ giving the relationship

$$s = 4.2t^{1.699}$$

[It is likely that the law $s = 4.2t^{1.7}$ would be a very good model for this data.]

Using regression on the TI-92

Alternatively we can use the TI-92 to go straight to a power law using the PowerReg calculation with $x = c5$ and $y = c6$. Figure 2.24 shows the two screens which lead directly to the power law

$$s = 4.20104t^{1.699238}$$

(remember $y = c6 = s$ and $x = c5 = t$).

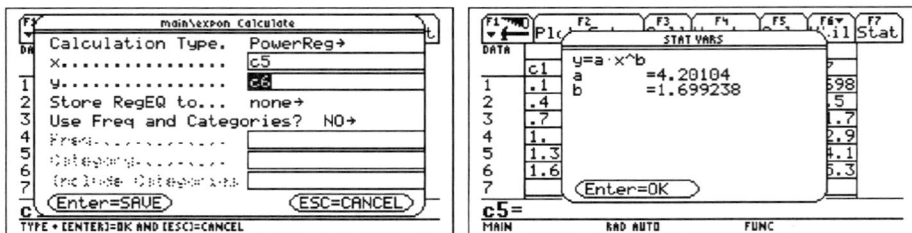

Figure 2.24   Using PowerReg to find $s = at^b$ directly

The TI-92 can be used to find an exponential law directly using the ExpReg utility. For the data for w and r in Table 2.7 use $x = c1$ and $y = c2$. Figure 2.25 shows the two screens which lead directly to the exponential law

$$w = 2.99 \times 0.018491^r$$

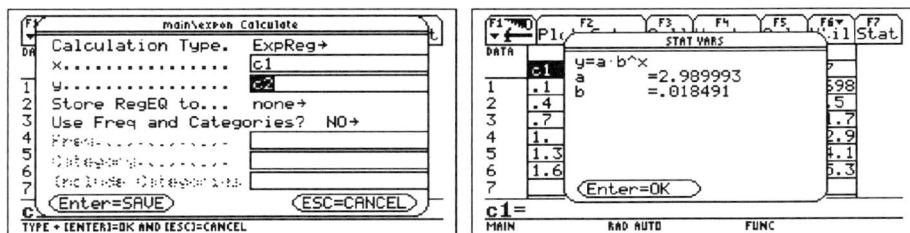

Figure 2.25   Using ExpReg to find $w = a.b^r$

Notice that the TI-92 does not give the model using the 'natural exponential' e. To recover the formula on page 81 we need to manipulate the base of the law, 0.018491.

$$w = a.b^r = a(e^{\ln(b)})^r = ae^{r\ln(b)}$$

(remember that $b = e^{\ln(b)}$.)

Putting $a = 2.99$ and $b = 0.018491$ we get

$$w = 2.99e^{-3.99r}$$

as before.

It is clearly important to interpret carefully the results of using ExpReg on the TI-92.

Exercise 2E

Use the TI-92 for each of problems 2-8.

1.   Which of the curves in Figure 2.26 suggests (a) a linear law, (b) a power law, (c) an exponential law, (d) none of these.

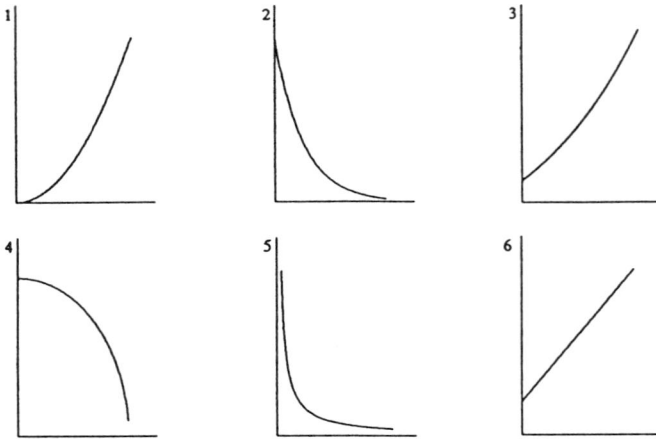

Figure 2.26

2.   Find the power law formula between the variables given in Table 2.8.

Table 2.8

(a)	x	1	2	3	4	5	6
	y	3.42	12.76	27.58	47.64	72.79	102.9

(b)	s	0.3	1.1	1.9	2.3	3.2	4.1
	t	16.7	1.24	0.42	0.28	0.15	0.089

3.   Find the exponential law formula between the variables given in Table 2.9.

Table 2.9

(a)	u	2	4	6	8	10
	v	2.68	1.80	1.20	0.808	0.541

(b)	x	0.1	0.3	0.7	0.9	1.6
	y	7.67	9.75	15.75	20.02	46.38

4.   The data in Table 2.10 shows the distance from the Sun and the period for each of the planets in the solar system.

Table 2.10

Planet	Distance, R (millions of km)	Period, T (days)
Mercury	57.9	88
Venus	108.2	225
Earth	149.6	365
Mars	227.9	687
Jupiter	778.3	4329
Saturn	1427	10753
Uranus	2870	30660
Neptune	4497	60150
Pluto	5907	90470

Find a relationship between T and R.

5.   Table 2.11 contains data that was produced in an experiment where the saturated vapour pressure of a fixed volume of water was determined at different temperatures.

Table 2.11

S.V.P. k (Nm^{-2})	0.61	0.86	1.21	1.70	2.33
Temperature T (°C)	0	5	10	15	20

Find a possible scientific model relating S.V.P. and temperature for this constant volume of water.

6.   Table 2.12 shows experimental values for the potential difference V and current I for a semiconductor diode.

Table 2.12

potential difference V (volts)	current I (microamps)
0.255	0.40
0.315	1.60
0.345	3.6
0.385	8.9
0.410	18.2
0.455	52.2
0.475	90.3
0.495	140
0.505	182
0.515	223
0.530	310

The exponential model $I = I_0 e^{aV}$ is proposed for this data. Show that this is a good model and find values of I_0 and a.

7.   The data in Table 2.13 gives the atmospheric pressure, expressed as a percentage of its sea-level value, at various altitudes.

Table 2.13

altitude, h (km)	0	5	10	14	20	24	30
pressure, p (% of sea level value)	100	53.0	26.0	14.0	5.4	2.9	1.2

Find a possible scientific model relating pressure and altitude.

8.   An experiment on the decomposition of nitrous oxide yielded the following values for the velocity constant k at various temperatures T.

Table 2.14

T(K)	985	1005	1058	1069	1105
k(mol min^{-1})	0.224	0.447	2.00	2.52	6.31

For most reactions a model between k and T is $k = ae^{-b/T}$ where a and b are constants. Show that this data fits such a model and find the values of a and b.

3

Trigonometric functions

3.1   INTRODUCTION

Oscillations occur frequently in real life, for example the swinging of a pendulum in a clock, the changing heights of tides, the movement of a needle in a sewing machine, and the amount of daylight as the seasons change. We will see that these oscillations can be modelled by the familiar sine and cosine functions.

3.2   DEGREES AND RADIANS

For most practical purposes angles are measured in degrees. Indeed, this has been the custom for about 4000 years, since the time of the Babylonian civilisation.

However, for the purposes of advanced work in mathematics another unit is needed. It is called the **radian**.

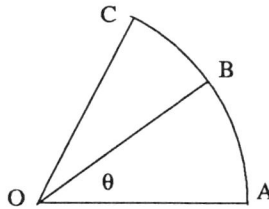

Figure 3.1

In Figure 3.1 the relative sizes of angles AOB and BOC may be compared by measuring the arc lengths AB and BC instead of counting the number of degrees turned through in each angle, provided $OA = OB = OC$. This is the essence of radian measure. Radians are arc lengths. In Figure 3.1 if arc length AB is equal to radial length OA then angle AOB is 1 radian. If the length of the arc AC is $2OA$ then the angle turned through is 2 radians, and so on.

In general the angle θ is given in radians by

$$\theta = \frac{\text{arc length } AB}{\text{radius } OA}$$

To find a connection between degrees and radians consider a circle of radius r.

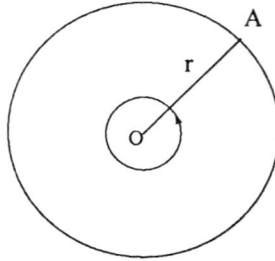

Figure 3.2

In one rotation of the radial line OA around the circle an angle of 360° is turned through. Since the radius of the circle is r, the distance travelled around the circle in one revolution is $2\pi r$. Then

$$\text{angle turned through} = \frac{\text{circumference}}{\text{radius } OA}$$

$$360° = \frac{2\pi r}{r}$$

Thus 2π radians = 360°, or π radians = 180°, or 1 radian = $180/\pi$ degrees (= 57.3°).
So, if angle AOB in Figure 3.3 is measured in degrees then

$$\theta \text{ degrees} = \frac{\pi\theta}{180} \text{ radians.}$$

On the other hand, if angle AOB is measured in radians then

$$\theta \text{ radians} = \frac{180\,\theta}{\pi} \text{ degrees.}$$

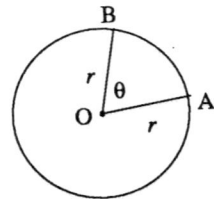

Figure 3.3

It is very common in advanced mathematics to state radians as multiples of π. For example,

$$90° = \frac{\pi}{180} \times 90 \text{ radians}$$

$$= \frac{\pi}{2} \text{ radians}$$

It is important when working with angles on your calculator to ensure that your calculator knows exactly which units you are using. On the TI-92 you can switch between degrees and radians using **MODE**, select Angle and then either RADIAN or DEGREE. The status line shows which you are working in.

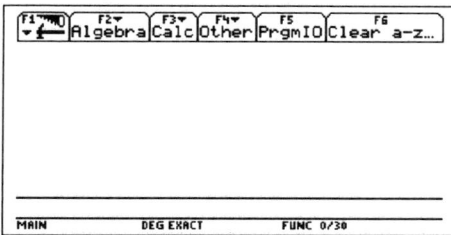

TI-92 in degree mode   TI-92 in radian mode

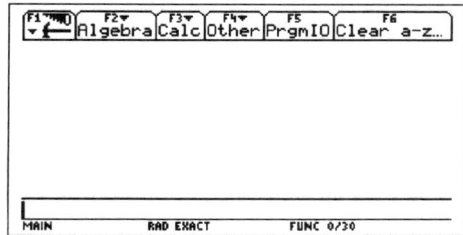

Consider again Figure 3.3; from the definitions of radians

$$\theta = \frac{\text{arc length } AB}{\text{radius}}$$

If the radius of the circle is r then we have the important result

> arc length $AB = r\theta$
> where θ is measured in radians.

The area of the circle is πr^2 and the sector OAB is a fraction of the circle given by

$$\frac{\text{area sector } OAB}{\text{area of circle}} = \frac{\theta}{2\pi}$$

Hence the area of the sector OAB is given by

area sector $OAB = \frac{1}{2}r^2\theta$

where θ is measured in radians.

Exercise 3A

1.   Use the conversion formulae given above and your calculator to change degrees into radians or radians into degrees in this problem.

(a) 65°   (b) 18°   (c) 1.5 radians   (d) 0.5 radians   (e) 2.8 radians

(f) 150°   (g) 5 radians   (h) 279°   (i) 1 radian   (j) 1°

2.   Complete the following table of degree and radian equivalence.

Table 3.1

Degrees	0	30	45	60	90	120	135	150	180
Radians	0		$\frac{\pi}{4}$		$\frac{\pi}{2}$				π

Degrees	210	225	240	270	300	315	330	360
Radians		$\frac{5\pi}{4}$			$\frac{5\pi}{3}$			2π

3.   Find in radians, the angle subtended at the centre of a circle of radius 20 cm by an arc:

(a)   8 cm long   (b)   35 cm long.

4.   Find in radians the angle of a sector of arc length 8 cm in a circle of radius 12 cm. What is the area of the sector?

5.   The area of a sector of a circle of radius 15 cm is 9 cm². Calculate length of the arc of the sector.

3.3   RIGHT ANGLED TRIANGLES

In elementary mathematics you will be familiar with the trigonometric ratios *sine*, *cosine* and *tangent* which are defined in terms of the sides of a right-angled triangle.

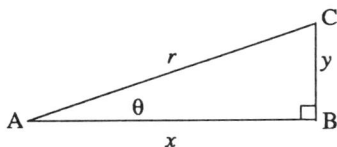

Figure 3.4

Thus for the right-angled triangle in Figure 3.4 we have

$$\sin\theta = \frac{y}{r} \qquad \cos\theta = \frac{x}{r} \qquad \tan\theta = \frac{y}{x}.$$

You will also be familiar with Pythagoras' theorem $x^2 + y^2 = r^2$.

TI-92 ACTIVITY 3A

The aims of this activity are (i) to show that having selected a particular unit for angles then you can still work in the other unit without changing MODE, and (ii) to use the TI-92 to solve for angles and sides of right angled triangles.

Clear the HOME screen and select APPROX mode.

(A)   (i)   Set up your TI-92 in RADIAN mode. To convert degrees to radians Type 45 **2nd D ENTER**. In exact mode this will give the answer $\pi/4$. In approx mode this will give 0.785398.

(ii)   Set up your TI-92 in DEGREE mode. To convert radians to degrees. Type ($\pi/4$) Select **MATH** (i.e. **2nd 5**), choose 2:Angle, 2:*r*, **ENTER**. In EXACT mode this gives 45. In APPROX mode any decimal places are included.

(B)   Use the TI-92 to convert degrees into radians or radians into degrees in this exercise.

(i)   82°   (ii)   3.72 radians   (iii) $\dfrac{7\pi}{5}$ radians

(iv)   250°   (v)   310°   (vi)   0.263 radians

(vii)   1 radian   (viii)   1°   (ix)   10.6°

(x) $\dfrac{8\pi}{15}$ radians

(C)   Figure 3.5 shows a ladder 3.5 m long leaning against a wall and making an angle of 61.3° with the ground. How far out from the wall is the foot of the ladder and how far up the wall is the top of the ladder?

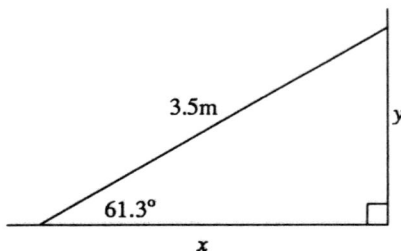

Figure 3.5

The length marked x denotes the distance of the foot of the ladder from the wall and the length marked y represents the height of the top of the ladder up the wall. To find x we recognise that the cosine function must be used.

$$\cos\theta = \frac{x}{r}$$

The TI-92 can solve this with one line of commands

Solve $(\cos\theta = (x/r), x)|\theta = 61.3°$ and $r = 3.5$

To find y we must use sine.

Solve $(\sin\theta = (y/r), y)|\theta = 61.3°$ and $r = 3.5$

θ can be found in [2nd, CHARACTER, GREEK, θ] or on the main keyboard. The screen dump in Figure 3.6 shows this activity.

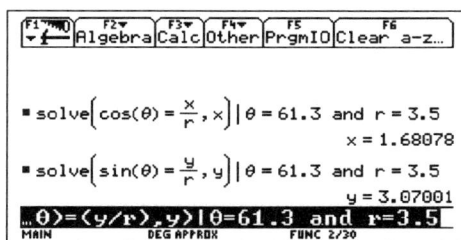

Figure 3.6 Solving for x and y

(D)   A rigid strut of length r metres is to support a bench 0.75 m wide. The strut is fastened to the end of the bench and 0.5 m below its contact with the wall, as shown in Figure 3.7. How long is the strut and what angle θ does it make with the wall?

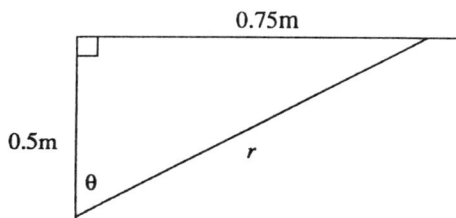

Figure 3.7

To find r we need to use Pythagoras' Theorem.
Set up your TI-92 in DEG APPROX mode.
Solve $(x^2 + y^2 = r^2, r) \mid x = 0.5$ and $y = 0.75$

gives $r = 0.901388$

Solve $(\tan(\theta) = y/x, \theta) \mid x = 0.5$ and $y = 0.75$.
This gives an answer of the form

Trigonometric equations have many solutions and the notation @n95 represents an arbitrary integer. (Your calculator may show a different value of j in the notation @nj; j can take values in the interval $1 - 255$)

$$\theta = 180(@n95 + 0.312833)$$

simply means $\theta = 180(n + 0.312833)$ for $n = 0, 1, 2, 3, 4, \ldots$

    $n = 0$ $\theta = 180(0.312833) = 56.3°$ (to 1D)
    $n = 1$ $\theta = 180(1 + 0.312833) = 236.3°$ (to 1D)
    $n = 2$ $\theta = 180(2 + 0.312833) = 416.3°$ (to 1D)

and so on.

It is obvious from our diagram of the problem that θ is less than $180°$ so if we take $n = 0$ this gives an answer $\theta = 56.3°$. Thus the strut needs to be 0.901m long and inclined at $56.3°$ to the wall.

Now solve the following problems using the TI-92.

(E)   A rectangular frame measures 9 m by 5 m and is kept in shape by two diagonal struts.

    (i)   Find the length of a strut.

    (ii)   What is the angle made by a strut with the longest side of the rectangle?

(F)   The road from a village P runs due east for 3 miles to a fort Q. There is a television mast due north of Q and 6 miles from Q. Find the distance and bearing of the mast from P.

(G)   A river runs parallel to a software plant. The roof of this plant is 35 m above the water level of the river. From a point on the roof, looking directly across the river, the angle of depression of the nearer bank is $39°$ and the angle of depression of the farther bank is $19°$. Find the width of the river.

(H)   In cartesian coordinates L is the point $(-2,-3)$, M is the point $(1,1)$ and N is the point $(2,-1)$. Find

    (i)   the angle between LM and a line parallel to the y axis,

    (ii)   the angle between MN and a line parallel to the x axis,

    (iii)   hence find angle LMN.

3.4   NON RIGHT ANGLED TRIANGLES

It is obvious that not all triangles are right angled and so it becomes necessary to develop further trigonometric techniques to deal with any sort of triangle.   Two important laws can be used in problems which do not involve right angled triangles; these are *the sine rule* and *the cosine rule*.

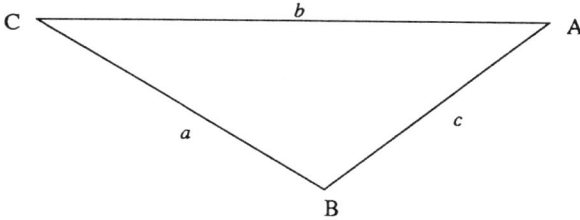

Figure 3.8

Figure 3.8 shows a helpful convention when dealing with non right angled triangles. The sides opposite the vertices A, B and C are designated a, b and c. Such a triangle is called **an oblique triangle**.

In Figure 3.9 the point D is the foot of the perpendicular line drawn from B to AC. The length of BD is h. If the length of AD is x then the length of CD will be $b - x$.

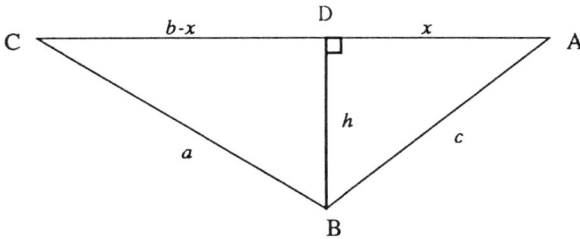

Figure 3.9

From triangles ABD and CBD,

$$\sin A = \frac{h}{c} \quad \text{and} \quad \sin C = \frac{h}{a}$$

so $c \sin A = h$ and $a \sin C = h$.

Then $c \sin A = a \sin C$

or $\qquad \dfrac{c}{\sin C} = \dfrac{a}{\sin A}$.

By choosing other altitudes and following the same argument we find that

$$\frac{b}{\sin B} = \frac{a}{\sin A} \quad \text{and} \quad \frac{c}{\sin C} = \frac{b}{\sin B}.$$

These three results may be drawn together into the **Sine Rule**.

$$\frac{a}{\sin A} = \frac{b}{\sin B} = \frac{c}{\sin C}$$

Now we develop the cosine rule. Consider Figure 3.9. Applying Pythagoras' Theorem to triangle ABD gives

$$h^2 = c^2 - x^2$$

and to triangle CBD,

$$h^2 = a^2 - (b-x)^2$$

hence

$$a^2 - (b-x)^2 = c^2 - x^2$$

$$a^2 - b^2 + 2bx - x^2 = c^2 - x^2$$

$$a^2 = b^2 + c^2 - 2bx$$

But from triangle ABD

$$\cos A = \frac{x}{c}$$

so

$$c \cos A = x$$

and therefore

$$a^2 = b^2 + c^2 - 2bc \cos A$$

In the same way, using other altitudes, it can be shown that

$$b^2 = a^2 + c^2 - 2ac \cos B$$

and $\quad c^2 = a^2 + b^2 - 2ab \cos C$

These last two results are of a similar form and lead to the **Cosine Rule**.

$$a^2 = b^2 + c^2 - 2bc \cos A$$
$$b^2 = a^2 + c^2 - 2ac \cos B$$
$$c^2 = a^2 + b^2 - 2ab \cos C$$

Furthermore, the area of triangle ABC is given by

$$area = \tfrac{1}{2}bh$$

but since $h = a \sin C$ we find

$$area = \tfrac{1}{2}ab \sin C$$

Since any altitude can be chosen, we find three versions of the area formula.

$$Area = \tfrac{1}{2}ab \sin C$$
$$= \tfrac{1}{2}bc \sin A$$
$$= \tfrac{1}{2}ac \sin B$$

Example 3A

Figure 3.10 shows a triangular field in which side AB is 400 m long, the angle at A is 65° and the angle at C is 100°.

(a)   What is the size of the angle at B?

(b)   How long are the sides BC and AC?

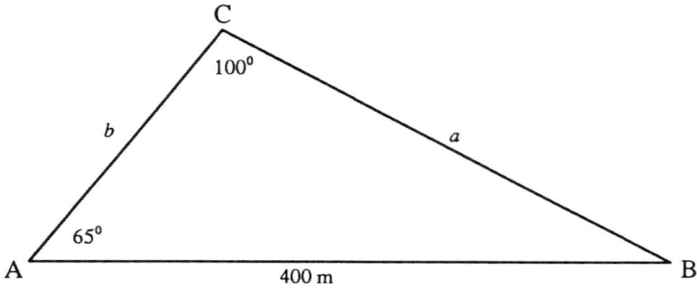

Figure 3.10

Solution

(a)   Since the angles in a triangle add up to 180°, the angle at B is 15°.

(b)   We use the sine rule in this problem to find a and b.

$$\frac{a}{\sin 65°} = \frac{b}{\sin 15°} = \frac{400}{\sin 100°}$$

Solving for a,

$$a = \frac{400\sin 65°}{\sin 100°} = 368.116 \text{ m}$$

Solving for b,

$$b = \frac{400\sin 15°}{\sin 100°} = 105.125 \text{ m}$$

The lengths of the two sides are 368 m and 105 m to 3 sf.

Example 3B

Figure 3.11 shows an oblique triangle in which angle B is 135° and sides a and c measure 3.15 cm and 2.72 cm respectively. Find the unknown angles and side length.

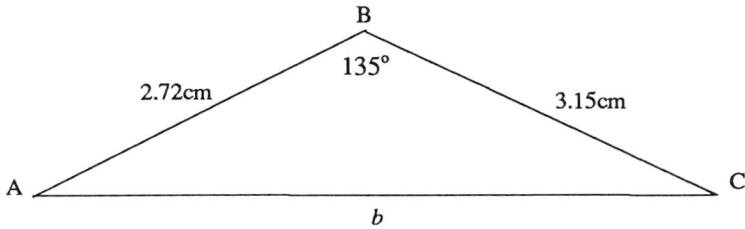

Figure 3.11

Solution

We use the cosine rule to find side length b.

$$b^2 = a^2 + c^2 - 2ac\cos B$$

$$b^2 = 3.15^2 + 2.72^2 - 2 \times 3.15 \times 2.72 \cos 135°$$
$$= 29.44$$

Hence

$$b = \sqrt{29.44} = 5.43 \text{ cm}$$

To find angle A (or C) we use the sine rule

$$\frac{a}{\sin A} = \frac{b}{\sin B}$$

Thus

$$\sin A = \frac{a \sin B}{b} = \frac{3.15 \sin 135°}{5.43} = 0.41$$

Solving for A we have

$$A = 24.2°$$

since the angles of a triangle add to 180°, the angle at C is 20.8°.

TI-92 ACTIVITY 3B

The aim of this activity is to show how the TI-92 can be used to do calculations like those in Examples 3A and 3B. Set the calculator to DEG and APPROX mode.

(A)   Consider again Example 3A. We must assign new names otherwise the TI-92 will treat Capital A the same as lower case a.
       The new notation will be

$$\frac{x}{\sin A} = \frac{y}{\sin B} = \frac{z}{\sin C}$$

in which $x \equiv a$, $y \equiv b$ and $z \equiv c$.

To solve for x (= a)

   Solve $(x/\sin(A) = z/\sin(C),x)\,|\,z = 400$ and $C = 100$ and $A = 65$

Press **ENTER** to give the answer $x = 368.116$ and to solve for y (= b)

   Solve$(y/\sin(B) = z/\sin(C),y)\,|\,z = 400$ and $C = 100$ and $B = 15$

giving $y = 105.125$.

(B)   Consider again Example 3B. To solve for b (or y) let $a = x$, $b = y$, $c = z$. Using the cosine rule $b^2 = a^2 + c^2 - 2ac \cos B$ becomes in our alternative notation

$$y^2 = x^2 + z^2 - 2xz \cos B$$

Type Solve $(y^2 = x^2 + z^2 - 2xz \cos B,y)\,|\,x = 3.15$ and $z = 2.72$ and $B = 135$ gives

   $y = 5.42566$

(We ignore the negative answer since a negative length has no meaning.) To find angle A use similar steps as in (A) to show that $A = 24.238°$.
       Now use the sine rule, cosine rule and the TI-92 to solve the following problems. Remember to use x, y, z instead of a, b, c when using the TI-92.

(C)   In triangle ABC, $A = 34.6°$, $C = 80.1°$ and $c = 10$ cm. Find a and b.

(D)   In triangle ABC, $a = 6.8$ cm, $b = 10.5$ cm and $B = 76.8°$. Find A and C and the area of the triangle.

(E)   A boat is sailing directly towards the foot of a cliff. The angle of elevation of a point on the top of the cliff, and directly ahead of the boat, increases from 6° to 10° as the boat sails 250 m. Find

     (i)   the original distance of the boat to the point on the top of the cliff,

     (ii)   the height of the cliff.

(F)   The side of a triangular field PQ is 150 m long. From P the bearing of the corner Q is 035° and the bearing of the corner R is 115°. From Q the bearing of R is 171°. Find

     (i)   the distances PR and QR,

     (ii)   the area of the field in m^2.

(G)   Find the angle between the hands of a clock at 10:20. If the hour hand is 15 cm long and the minute hand is 20 cm long, how far apart are the tips of the hands at this time?

(H)   A ship leaves port A and sails 12 nautical miles on a bearing of 025° followed by 15 nautical miles on a bearing of 192° to arrive at the next port B. Find the distance and bearing of A from B.

3.5   THE COSINE AND SINE FUNCTIONS

Trigonometry is a very powerful branch of mathematics. This becomes most evident once it is seen to break free of its apparent confinement to triangles. Anything that *oscillates*, for example a pendulum swinging or the variation of daylight with the changing seasons, may be *modelled* mathematically by means of the *cosine* and *sine* functions.

    Consider a circle of unit radius called *the unit circle* with its centre at the origin of a Cartesian coordinate system (Figure 3.12).

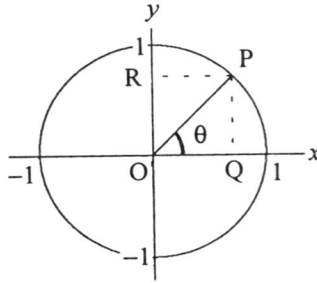

Figure 3.12

A point P on the circumference of this circle makes an angle θ with the positive direction of the x axis. For simplicity, P is shown lying in the first quadrant.

Right angled triangles OPQ and OPR are formed with the coordinate axes and since $OP = 1$, the lengths of OQ and OR are given by

$$OQ = \cos \theta \quad \text{and} \quad OR = \sin \theta$$

If we rotate the point P in an anticlockwise direction, so that θ increases, then we can define

$OQ = x$ coordinate of the point $P = \cos \theta$
$OR = y$ coordinate of the point $P = \sin \theta$

With these definitions the values of sine and cosine can be found for any angle. The graphs of the x and y coordinates of P are shown in Figures 3.13 and 3.14. (Note that a clockwise rotation of P corresponds to negative values of θ).

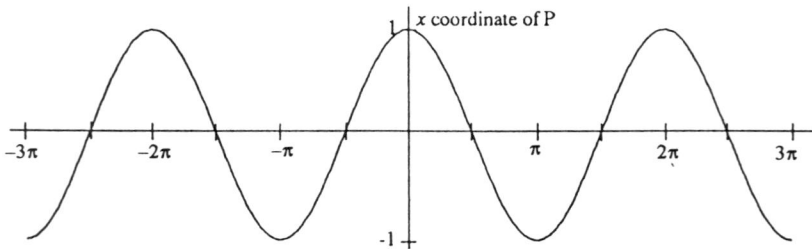

Figure 3.13   Graph of x coordinate of $P = OQ$

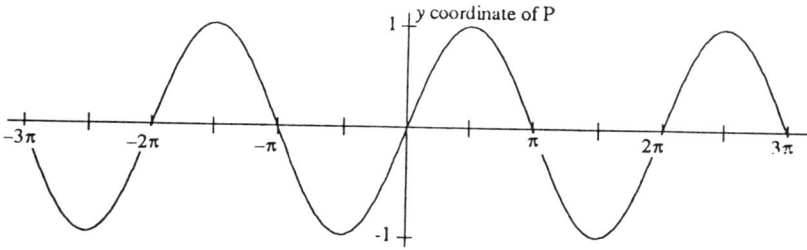

Figure 3.14 Graph of y coordinate of $P = OR$

Figure 3.13 shows the graph of **the cosine function** for angles of any magnitude while Figure 3.14 shows the graph of **the sine function**. From these graphs we can deduce important features of the sine and cosine functions.

1.   The graphs of both functions repeat themselves every 2π radians: they are **periodic** with a **period** of 2π

$$\cos(\theta + 2\pi) = \cos\theta \qquad \sin(\theta + 2\pi) = \sin\theta.$$

2.   The graph of the cosine function has line symmetry about its vertical axis: it is an example of an **even function**

$$\cos(-\theta) = \cos\theta$$

3.   The graph of the sine function has a half-turn rotational symmetry about the origin: it is an example of an **odd function**

$$\sin(-\theta) = -\sin\theta$$

4.   Both graphs have a range that is limited: they are **bounded** between +1 and −1 inclusive

$$-1 \le \cos\theta \le 1 \qquad -1 \le \sin\theta \le 1.$$

5.   Both graphs have no breaks in them and their domains are unlimited: they are **continuous**.

6.   If the graph of the cosine function is translated to the right by $\dfrac{\pi}{2}$ radians it becomes the graph of the sine function

$$\cos\left(\theta - \frac{\pi}{2}\right) = \sin\theta$$

7.   If the graph of the sine function is translated to the left by $\dfrac{\pi}{2}$ radians it becomes the graph of the cosine function

$$\sin\left(\theta + \frac{\pi}{2}\right) = \cos\theta$$

Exercise 3B

1.   Select RAD and APPROX mode. Define WINDOW to be

   xmin $= -2\pi$,   xmax $= 4\pi$,   xscl $= \pi/2$,
   ymin $= -2$,   ymax $= 2$,   yscl $= 1$

   (Note that when you Input them as above, the TI-92 converts them to Approx representation.)
   Graph the sine and cosine curves for the range of values -2π to $+4\pi$. Check properties 1 to 7 above using your graphs.

2.   Plot the graphs of the following functions.

   (a) $y = x$ (b) $y = x^2$ (c) $y = x^3$

   (d) $y = e^x$ (e) $y = e^{-x}$ (f) $y = 2$

   (g) $y = x\sin x$ (h) $y = x\cos x$ (i) $y = x^2\cos x$

   (j) $y = \dfrac{1}{1+x^2}$ (k) $y = e^{-x^2}$

   Classify each function as even, odd, or neither even nor odd.

3.6   THE TANGENT FUNCTION

From elementary trigonometry you will know that

$$\tan\theta = \frac{\sin\theta}{\cos\theta}$$

Now we can extend the graph of $\tan\theta$ beyond the domain $0 \le \theta \le \dfrac{\pi}{2}$ with which you are familiar.   Figure 3.15 shows the graph of $\tan\theta$ for the domain $-3\pi \le \theta \le 3\pi$.

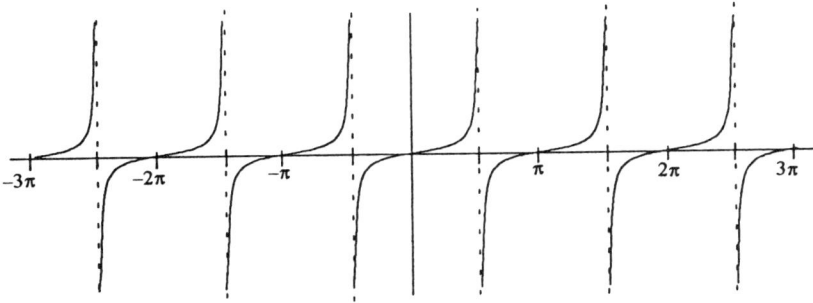

Figure 3.15   Graph of tan(θ)

Exercise 3C

1.   Use Figure 3.15 to deduce the properties of the tangent function (compare properties 1 to 5 for the sine and cosine functions).

2.   Although $\tan\theta$ is not defined for $\theta = 90°$ it is defined for values of θ close to 90°.

Find the values of

(a)   tan 89°   (b)   tan 89.9°   (c)   tan 89.99°
(d)   tan 90.01°   (e)   tan 90.1°   (f)   tan 91°

3.   At $\theta = 90°$ the graph of $\tan\theta$ has an asymptote.   Where else does $\tan\theta$ have asymptotes?

3.7   FEATURES OF TRIGONOMETRIC FUNCTIONS

TI-92 ACTIVITY 3C

The aim of this activity is to explore the properties of the trigonometric functions and their graphs. Set your TI-92 to RAD and APPROX mode.

(A)   (i)   Change the range for y to $y$min = –4 and $y$max = 4. Return to the HOME screen.

Type Graph{1,2,3}sin(x) **ENTER**

This generates the graphs sin(x), 2sin(x), 3sin(x). Figure 3.16(a) shows the graphs of the three sine functions. How do the three graphs compare?

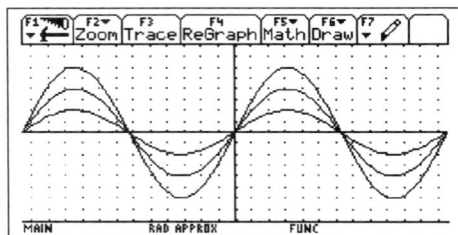 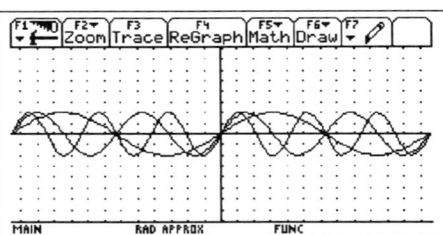

Figure 3.16(a)   Figure 3.16(b)

(ii)   Repeat for cos x.

(iii)   Repeat for tan x.

(B)   (i)   Return to the HOME screen.

Type Graph sin({1,2,3}x) **ENTER**

This generates sin$x$, sin2x, sin3x. Figure 3.16(b) shows the graphs of the three sine functions. How do the three graphs compare?
Delete the graphs.

(ii)   Repeat for cos x.

(iii)   Repeat for tan x.

(iv)   Try sketching by hand a graph of the function 3sin2x.   Check your sketch by plotting the function on the TI-92.

(v)   Repeat (iv) with other functions of the form $m$sin(nx) for different values of m and n.

(C)   (i)   Graph sin(x + {0,0.5,1,1.5}).
In which direction are successive graphs being translated? **Zoom In** on the values to the left of the origin where successive graphs cross the x axis. How do these values relate to the numbers you see in the vector?

Reset the WINDOW, delete the graphs and repeat the procedure for Graph (sin(x − {0,0.5,1,1.5})). **Zoom In** on the values to the right of the origin where successive graphs cross the x axis.

(ii)   Repeat for Graph cos(x − {0,0.5,1,1.5}). **Zoom In** on the peaks to the left and right of the origin.

(iii)   Repeat for Graph tan(x + {0,0.5,1,1.5}), Graph tan(x − {0,0.5,1,1.5}).

(iv)   Graph sin(2x + {0,0.5,1,1.5}).   Graph sin(4x − {0,0.5,1,1.5}).

(D)   (i)   Graph sin(x) + {0,1,2,3}.   In which direction are the graphs translated? **Zoom In** on those values where successive graphs cross the y axis. How do these values relate to the numbers in the vector?   Reset the WINDOW, delete the graphs and repeat the procedure for Graph sin(x) − {0,1,2,3}.

(ii)   Repeat with cosine.

(iii)   Repeat with tangent.

These four activities have shown that the graphs of the trigonometric functions sine, cosine and tangent may be subtly altered by including numbers in various places within the function description.   Particularly important are the sine and cosine functions which are often used to model physical situations involving oscillations.

Consider the sine function of time t

$$f(t) = a\sin(wt + \alpha) + c$$

which may be regarded as typical of them all.

The four parameters, a, w, α and c each cause a specific alteration in the graph of $\sin t$. Taking them in the order of the investigations

1. a is the **amplitude** of the function and affects the size of the oscillations.

2. w is the **angular frequency** of the function and affects the number of oscillations.

3. α is related to a **horizontal translation** which is determined by solving for t the equation $wt + \alpha = 0$ i.e. $t = -\dfrac{\alpha}{w} = t_0$; the graph is translated to the left if $\alpha > 0$ and to the right if $\alpha < 0$.

4. c is a **vertical translation** which is upwards if $c > 0$ and downwards if $c < 0$.

There are two further related concepts to consider. We saw in section 3.5 that the period of the fundamental sine function is 2π. Since the function $\sin 2t$ has twice as many oscillations, its period must be $2\pi/2 = \pi$. Likewise the function $\sin 3t$ has three times as many oscillations and so its period must be $2\pi/3$.

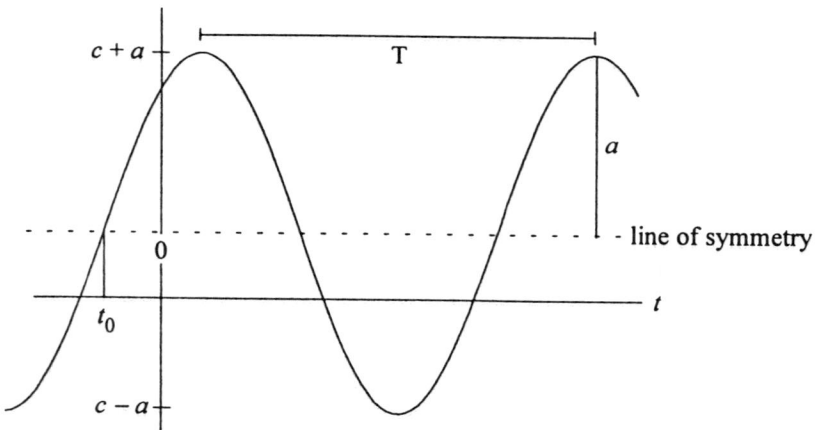

Figure 3.17   Properties of $a\sin(wt + \alpha) + c$

In general the **period**, T, of a trigonometric function with angular frequency w is given by $T = \dfrac{2\pi}{w}$. Thus we see that the higher the angular frequency the shorter is the period.

The word "period" suggests an interval of time (which is why we have written f in terms of t) and is appropriate to functions which repeat themselves in cycles. It can be important to know how much of the cycle takes place in a given unit of time. This is known as **frequency**, and in practical circumstances is typically measured in cycles per second or Hertz (Hz).

In general the frequency, f, of a trigonometric function of period T will be given by $f = \dfrac{1}{T}$. Thus

$$f = \frac{w}{2\pi}$$

or $w = 2\pi f$

which shows that the angular frequency is directly proportional to the frequency.

▓▓▓▓▓▓▓▓▓ **Exercise 3D** ▓▓▓▓▓▓▓▓▓

1.   State (i) the amplitude, (ii) the angular frequency, (iii) the horizontal translation and (iv) the vertical translations of the following expressions.

(a) $3\sin 5t$

(b) $4\cos\left(t - \dfrac{\pi}{2}\right)$

(c) $\cos(3t + 2\pi)$

(d) $\sin(3t) + 2$

(e) $\sin(3t + 2)$

(f) $2\sin(4t - \pi) + 1$

(g) $\cos(\tfrac{1}{2}t - \pi) - 3$

(h) $0.1\sin(2\pi t - 3\pi) + 0.5$

2.   Sketch (by hand) the graphs of the expressions in problem 1. (Check your answers using the TI-92).

3.   Write down:

(a)   the cosine function with amplitude 6 and angular frequency 2

(b)   the sine function with amplitude 3, angular frequency 4 and horizontal translation $-\dfrac{\pi}{4}$

(c)   a tangent function with angular frequency 5 and vertical translation 3

(d)   the sine function with amplitude 10, angular frequency 5 and horizontal translation $\dfrac{\pi}{10}$

(e)   the cosine function with amplitude 0.5, angular frequency 3, horizontal translation $\dfrac{2}{3}$ and vertical translation -1

(f)   the period of a trigonometric function with angular frequency 20

(g)   the frequency of a trigonometric function with period 0.1π

(h)   the angular frequency of a trigonometric function with frequency 100 Hz.

4.   Use the TI-92 to plot the functions (a) to (e) in problem 3.

5.   State (i) the period, (ii) the frequency of the following functions.

(a) $\cos(3t-1)$ (b) $\sin(5t+\pi)$ (c) $4\sin(8\pi t+1)$

(d) $10\cos(0.5\pi t-3\pi)$ (e) $\sin\left(8t-\dfrac{\pi}{2}\right)$

6.   Sketch (by hand) the graphs of the functions in problem 5. (Check your answers using the TI-92.)

7.   For each of the following expressions (i) use of an appropriate table of values to explore the limit as $x \to 0$, (ii) draw a graph of the function, (iii) find the limit as $x \to 0$ using the TI-92.

(a) $\dfrac{x}{\sin x}$ (b) $\dfrac{\sin x}{x^2}$

(c) $\dfrac{1-\cos x}{x}$ (d) $\dfrac{\sin x - x\cos x}{x^3}$

3.8   MODELLING WITH TRIGONOMETRIC FUNCTIONS

In this section we shall apply the concepts of amplitude, frequency, period, etc. to physical phenomena.

Example 3C

The following table represents the number of hours of daylight in the city of Plymouth over two years. The measurements began on 15th April 1990. Express the number of hours of daylight N as a function of the number of days t since 15th April 1990.

t	0	91.25	182.5	273.75	365	456.25	547.5	638.75	730
N	14	19	14	9	14	19	14	9	14

Solution

We are looking for a function of the form $N = a\sin(wt + \dot{\alpha}) + c$. We see that the largest and smallest values of N are $N = 9$ and $N = 19$, so the amplitude of the oscillation $a = \frac{1}{2}(19 - 9) = 5$ and the vertical translation $c = 14$.

The period of the phenomenon is 365 days which gives an angular frequency of $w = \dfrac{2\pi}{365} = 0.0172$. The horizontal shift is zero because when $t = 0$, $N = 14$ the mean value of the oscillation which gives

$$5\sin\alpha + 14 = 14$$
$$5\sin\alpha = 0$$

hence

$$\alpha = 0$$

Thus the model for the number of daylight hours as a function of time t is

$$N = 5\sin(0.0172t) + 14$$

Figure 3.18 shows the screen dump of the data and the model. We see that the fit is very good.

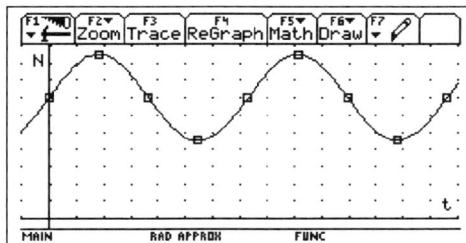

Figure 3.18 Data and sine model for Example 3C

▒▒▒▒▒▒▒▒▒▒▒▒▒▒ **Example 3D** ▒▒▒▒▒▒▒▒▒▒▒▒▒▒▒▒▒▒▒▒▒▒▒▒▒▒▒▒▒▒▒▒▒▒▒▒

The blood pressure, P millibars (mb), of a patient in hospital is modelled by the function

$$P = 25\cos(6t) + 95$$

where t is time measured in seconds.

Use this model to find the maximum (systolic) and minimum (diastolic) pressures. What is the period of time between consecutive systolic pressures? What is the patient's blood pressure after 2 seconds?

Solution

The amplitude of the function, $a = 25$, will be attained whenever $\cos(6t) = 1$ or $\cos(6t) = -1$.

Thus the systolic pressure is $25 + 95 = 120$ mb and the diastolic pressure is $-25 + 95 = 70$mb.

Consecutive maxima (or minima) are repeated in a time equal to the period of the oscillation i.e. $\dfrac{2\pi}{6} = 1.047$ s.

When $t = 2s$ (remembering to work in radians) the pressure is given by

$$P = 25\cos(12) + 95 = 116.096 \text{ mb (to 3 dp)}.$$

▒▒

These examples show how we can use the functions sine and cosine to model any system which behaves in a wave like motion. All that is required is either a graph showing the motion of the system or data concerning the period, amplitude and frequency. The functions sine and cosine are often called **sinusoidal functions**.

Exercise 3E

1. In the USA, electricity is supplied at a frequency of 60 Hz. What function would model the current across a resistor if the current's amplitude was 80 mA?

2. A mass attached to the end of a spring oscillates so that its displacement s cm from a central position as given by

    $$s = 6\sin(3\pi t)$$

    where t is the time in seconds. What is the maximum displacement from the central position? What is the frequency of the oscillation? Where is the object

    (a) 5 seconds into the motion?

    (b) 17.3 seconds into the motion?

3. The mean monthly temperature in Benidorm peaks in August at 27°C and has a minimum in February of 5°C. Assuming that the variation in temperature is sinusoidal, obtain a mathematical model to represent the mean monthly temperature. Use this model to predict the mean monthly temperature in June and in January.

4. The Bristol Channel is known to have an unusually high rise and fall in tide levels. For a typical spring tide there is as much as a 13m difference between high tide and low tide. Given that there is a gap of 12.4 hours between successive high tides, find the values of a and w in the model

    $$h = a\sin(wt+\alpha)$$

    which gives the height h in metres of the water above mean sea level as a function of time t in hours.

    On a given day, at the time of a spring tide, high water occurs at midnight ($t = 0$). Find the value of α and use the model to determine the height above mean sea level of the water at midnight on the following three successive nights.

3.9   SOLVING TRIGONOMETRIC EQUATIONS

In this section we shall see that solving trigonometric equations is a quite different process to solving polynomial equations.

TI-92 ACTIVITY 3D

(A)   (i)   Draw the graph of sin x for $-\pi \le x \le 3\pi$.

      (ii)   Draw the line $y = 0.4$. How many intersection points of the two graphs can you see? How many more intersection points do you think there are?

      (iii)   Use Trace to find some of the solutions of the equation sin $x = 0.4$.

(B)   Repeat (A) using cos x.

(C)   Repeat (A) using tan x.

From this activity we can see that even the simplest kinds of trigonometric equation,

$$\sin x = 0.4, \quad \cos x = 0.4, \quad \tan x = 0.4$$

possesses infinitely many solutions. This is because the trigonometric functions sin, cos and tan are periodic.

Exercise 3F

Use your TI-92 or other calculator in the degree mode to solve the following using SIN^{-1}, COS^{-1} or TAN^{-1}.

1.	sin $x = 0.4$	2.	cos $x = 0.4$	3.	tan $x = 0.4$
4.	sin $x = -0.8$	5.	cos $x = -0.8$	6.	tan $x = -0.8$.

      The results obtained in Exercise 3F show that an electronic calculator will supply only *one* of the infinitely many possible answers to each problem. The reason for this will be explained more fully in section 3.10.

Figure 3.19 shows the graphs of $y = \sin x$ and $y = 0.4$ for $-540° \le x \le 540°$.

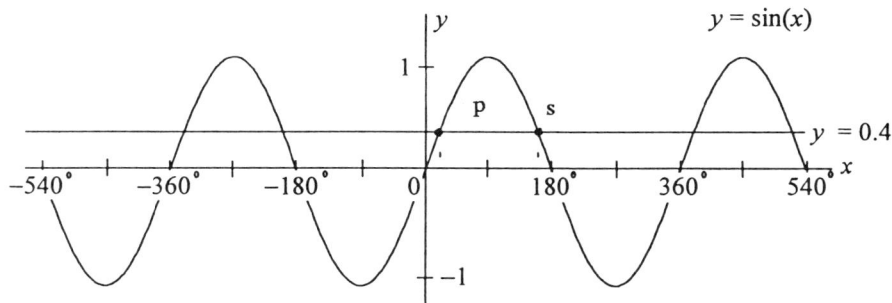

Figure 3.19

When solving the equation sin x = 0.4 with a calculator the answer x = 23.6° is obtained.   In Figure 3.19 this solution is labelled p because we shall call it the **primary solution** of the equation.

Because of the symmetry of the sine graph the next solution nearest to p is 180 − 23.6° = 156.4°.   In Figure 3.19 this is labelled s because we shall call it the **secondary solution** of the equation.

These solutions, p and s, together with the periodic nature of the sine graph will supply all possible solutions to the equation sin x = 0.4. Since we know that the period of the sine graph is 360° we can now see that the general solution of the equation sin x = 0.4 must be

$$x = 360°n + 23.6°$$

or

$$x = 360°n + 156.4°$$

for n = 0,±1,±2,±3,.....

On the TI-92 we can obtain these solutions directly using solve. Figure 3.20 shows screen dump for the problem of solving sin(x) = 0.4. Expanding the brackets gives

$$x = 360@n88 + 156.422 \quad \text{or} \quad x = 360@n88 + 23.5782$$

As we saw on page 93, the symbol @n88 represents an arbitrary integer n. (Your calculator may show a different digit after @n, this does not matter.)

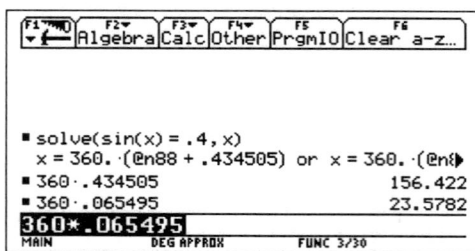

Figure 3.20   Solving sin $x = 0.4$

▒▒▒▒▒▒▒▒▒▒▒ **Example 3E** ▒▒▒▒▒▒▒▒▒▒▒▒▒▒▒▒▒▒▒▒▒▒▒▒▒▒▒▒▒▒▒▒▒

Solve sin $x = -0.73$ for $-720° \le x \le 720°$.

Solution

Using a calculator we find the primary solution $p = -46.9°$. The secondary solution is given by $s = -180° + 46.9° = -133.1°$. So the general solution is given by

$$x = 360°n - 46.9° \quad \text{or} \quad x = 360°n - 133.1°$$

Taking $n = 0$ we have the primary and secondary solutions

$$x = -46.9° \quad \text{and} \quad x = -133.1°$$

Taking $n = 1$ we obtain

$$x = 313.1° \quad \text{and} \quad x = 226.9°$$

Taking $n = -1$ we obtain

$$x = -406.9° \quad \text{and} \quad x = -493.1°$$

Taking $n = 2$ we obtain

$$x = 673.1° \quad \text{and} \quad x = 586.9°$$

Taking any more values of n would take us outside the required interval $-720° \le x \le 720°$. Thus the solutions in this interval are

$$x = -493.1°, -406.9°, -133.1°, -46.9°, 226.9°, 313.1°, 586.9°, 673.1°$$

Figure 3.21 shows some of these solutions graphically on a TI-92 screen.

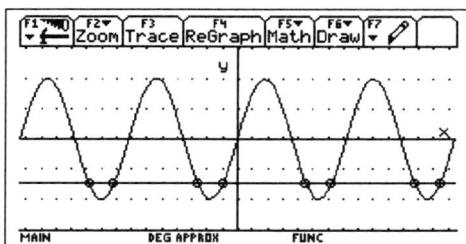

Figure 3.21   Solutions of sin $x = -0.73$

A similar procedure may be adopted for equations involving cos and tan.

Example 3F

Solve $\cos x = 0.4$ for $-360° \leq x \leq 360°$.

Solution

The primary solution given by a calculator is $p = 66.4°$.
    In the case of the cosine graph the secondary solution is $s = -66.4°$ as shown in Figure 3.22.

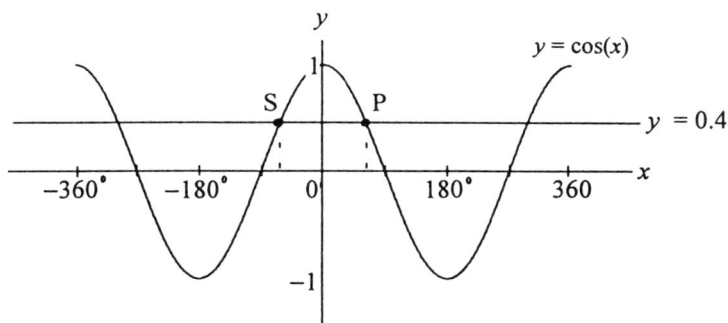

Figure 3.22

The period of the cosine graph is 360° and therefore we see with reference to Figure 3.21 that the general solution of $\cos x = 0.4$ must be

$$x = 360°n \pm 66.4°$$

When $n = 0$ we get $x = 66.4°$ and $-66.4°$.
When $n = 1$ we get $x = 293.6°$ and $-293.6°$.

Any further values of n will take us outside the range $-360° \leq x \leq 360°$.

So the solutions in this interval are

$$x = -293.6°, -66.4°, 66.4°, 293.6°.$$

Example 3G

Solve $\tan x = -0.8$ for $-450° \leq x \leq 450°$.

Solution

The primary solution $p = -38.7°$. Figure 3.23 shows that there is no secondary solution for simple tangent equations.

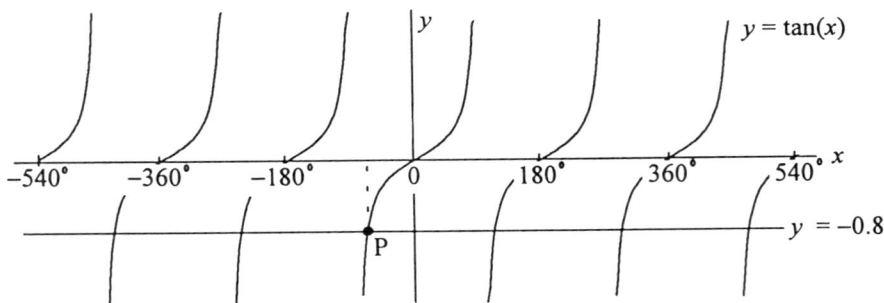

Figure 3.23

The period of the tangent function is $180°$ and we can see from Figure 3.23 that the general solution of $\tan x = -0.8$ is

$$x = 180°n + (-38.7°)$$

Thus

when $n =$		
when $n =$	0,	$x = -38.7°$
when $n =$	1,	$x = 141.3°$
when $n =$	-1,	$x = -218.7°$
when $n =$	2,	$x = 321.3°$
when $n =$	-2,	$x = -398.7°$
when $n =$	3,	$x = 501.3°$

Any further value of n takes us outside the interval $-540° \le x \le 540°$. So we have the required solutions

$$x = -398.7°, -218.7°, -38.7°, 141.3°, 321.3°, 501.3°.$$

Exercise 3G

1.   For $-360° \le x \le 360°$ solve the equations

    (a) $\sin x = 0.47$ (b) $\cos x = -0.17$ (c) $\tan x = 1.4$

    (d) $\cos x = 0.68$ (e) $\tan x = -2$ (f) $\sin x = -0.89$

2.   Use the TI-92 to display your answers to problem 1 graphically.

3.   For $-180° \le \theta \le 540°$ solve the equations

    (a) $\cos \theta = 0.31$ (b) $\tan \theta = -0.89$ (c) $\sin \theta = 0.9.$

4.   Use the TI-92 to display your answers to problem 2 graphically.

5.   Use the **F2** 1:solve command on the TI-92 to find all solutions of the following equations for $-2\pi \le t \le 2\pi$ Give your answers in radians.

    1. $\cos t = 0.47$ 2. $\sin t = 0.163$ 3. $\tan t = -0.09$

    4. $\sin t = -0.14$ 5. $\tan t = 1.3$ 6. $\cos t = 0.01$

3.10   INVERSE TRIGONOMETRIC FUNCTIONS

Section 3.9 showed us that trigonometric equations have infinitely many solutions and you saw how to state general solutions for simple sine equations, cosine equations and tangent equations.

We noted that our electronic calculators supply only one of the infinitely many possible solutions, and we called this the primary solution. Graphical considerations enabled us to find a secondary solution and then an expression for a general solution. The TI-92 gives the general solution to an equation involving trigonometric functions.

It would be impractical for your calculator or computer to deliver infinitely many answers to an equation. When we find the *inverse* of a trigonometric function, using SIN^{-1} for example, it is necessary to restrict the range of possible values so that only one value, usually the smallest or most convenient, is selected. We can then determine what the other solutions will be through the methods demonstrated in section 3.9. These restricted ranges of possible values are called **principal values** and they provide the primary solution to any trigonometric equation.

Notations for the inverse trigonometric functions vary. The most usual are

$$y = \sin^{-1} x \qquad y = \arcsin x \qquad y = \text{ASIN } x$$
$$y = \cos^{-1} x \qquad y = \arccos x \qquad y = \text{ACOS } x$$
$$y = \tan^{-1} x \qquad y = \arctan x \qquad y = \text{ATAN } x$$

Figure 3.24, Figure 3.25 and Figure 3.26 show the graphs of the inverse trigonometric functions and their principal values.

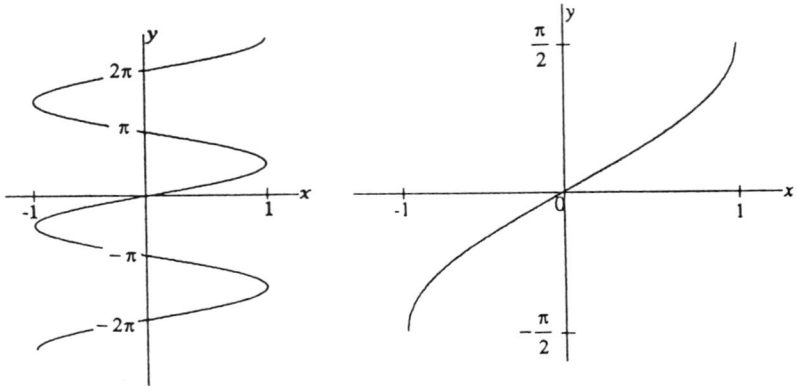

Figure 3.24   Graph of $y = \arcsin x$

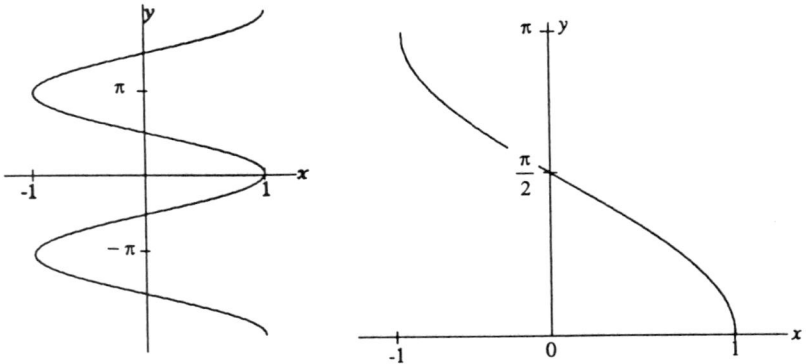

Figure 3.25   Graph of $y = \arccos x$

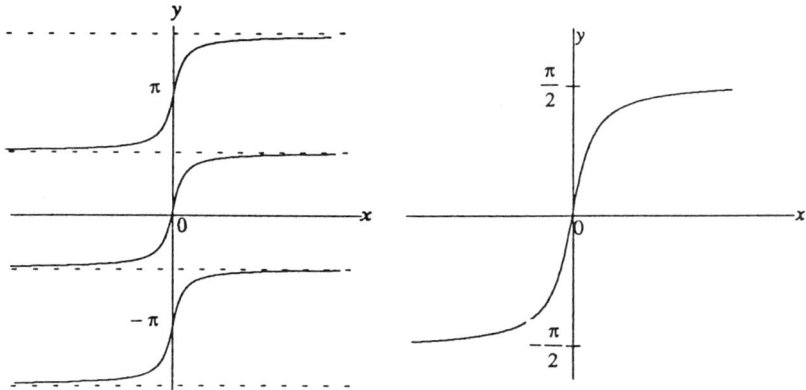

Figure 3.26   Graph of arctan x

Exercise 3H

1. Using your calculator find the principal value in degrees of each of the following.

    (a) $\cos^{-1}(0.62)$ (b) $\tan^{-1}(-1.05)$ (c) $\sin^{-1}(-0.88)$

    (d) $\tan^{-1}(6.59)$ (e) $\sin^{-1}(0.06)$ (f) $\cos^{-1}(-0.95)$

2. Use your calculator to find the principal value in radians of each of the following.

    (a)   ATAN(−3.5)   (b)   ASIN(−0.1)   (c)   ACOS(0.06)

    (d)   ACOS(−0.33)   (e)   ATAN(0.5)   (f)   ASIN(0.11)

3. Find the values of each of the following expressions.

    (a) $\sin(\cos^{-1}0.1)$ (b) $\cos(\sin^{-1}0.1)$ (c) $\tan(\cos^{-1}0.8)$

    (d) $\tan(\sin^{-1}-0.3)$ (e) $\cos(\tan^{-1}3.2)$ (f) $\sin(\tan^{-1}-1.6)$

    (g) $\sin(\sin^{-1}0.8)$ (h) $\cos(\cos^{-1}-0.55)$ (i) $\tan(\tan^{-1}0.75)$

3.11   TRIGONOMETRIC IDENTITIES

You will have already seen that there are some simple relationships between sine, cosine and tangent. Often when solving a problem it can be necessary to change the form of a trigonometric expression in order to progress. In this section we will consider a number of other important relationships often known as **identities**.

The first result, sometimes referred to as the **Pythagorean Relationship**, can be deduced by considering the right angle triangle shown in Figure 3.27. We know that

$$a^2 + b^2 = c^2$$

which can be rewritten as

$$\frac{a^2}{c^2} + \frac{b}{c^2} = 1$$

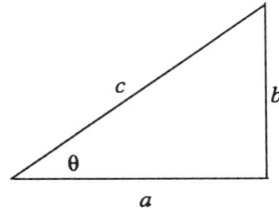

Figure 3.27

or

$$\left(\frac{a}{c}\right)^2 + \left(\frac{b}{c}\right)^2 = 1$$

But from Figure 3.23 $\sin\theta = \dfrac{b}{c}$ and $\cos\theta = \dfrac{a}{c}$, so the equation above becomes

$$(\sin\theta)^2 + (\cos\theta)^2 = 1$$

which is usually written as

$$\sin^2\theta + \cos^2\theta = 1$$

This is a very important result, known as an **identity**, and one that you will encounter frequently in your studies.

It is also useful to introduce three new functions which are defined as the reciprocals of the basic three trigonometric functions. They are

secant $\sec\theta = \dfrac{1}{\cos\theta}$ **cosecant** $\operatorname{cosec}\theta = \dfrac{1}{\sin\theta}$

cotangent $\cot\theta = \dfrac{1}{\tan\theta} = \dfrac{\cos\theta}{\sin\theta}$

With these new functions and the Pythagorean relationship we can solve more complicated looking trigonometric equations.

Example 3H

Find the solution of the equation

$$\cos\theta - \sec\theta = \tan\theta\ ,$$

for $0 \le \theta < 360°$.

Solution

The first step is to express sec and tan in terms of sin and cos to give

$$\cos\theta - \frac{1}{\cos\theta} = \frac{\sin\theta}{\cos\theta}$$

Now multiplying by $\cos\theta$ gives

$$\cos^2\theta - 1 = \sin\theta$$

Using $\sin^2 + \cos^2\theta = 1$ (in the form $\cos^2\theta = 1 - \sin^2\theta$) leads to

$$1 - \sin^2\theta - 1 = \sin\theta$$

or

$$\sin\theta + \sin^2\theta = 0$$

Factorising gives

$$\sin\theta(1 + \sin\theta) = 0$$

so

$$\sin\theta = 0 \quad \text{or} \quad \sin\theta = -1$$

so

$$\theta = 0°,\ 180° \text{ or } 270°.$$

It is possible to form other identities in terms of sec, cosec and cot from the identity obtained above. For example starting with

$$\sin^2\theta + \cos^2\theta = 1$$

we can divide all terms by $\cos^2\theta$ to obtain

$$\frac{\sin^2\theta}{\cos^2\theta} + \frac{\cos^2\theta}{\cos^2\theta} = \frac{1}{\cos^2\theta}$$

or

$$\tan^2\theta + 1 = \sec^2\theta$$

Example 3I

Prove that

$$\tan^2\theta - \sin^2\theta = \tan^2\theta \sin^2\theta$$

Solution

Starting with the LHS and expression $\tan\theta$ in terms of $\sin\theta$ and $\cos\theta$ gives

$$\tan^2\theta - \sin^2\theta = \frac{\sin^2\theta}{\cos^2\theta} - \sin^2\theta$$
$$= \sin^2\theta\left(\frac{1}{\cos^2\theta} - 1\right)$$
$$= \sin^2\theta(\sec^2\theta - 1)$$

But we know the identity $\tan^2\theta + 1 = \sec^2\theta$, so $\tan^2\theta = \sec^2\theta - 1$. Using this gives

$$\sin^2\theta(\sec^2\theta - 1) = \sin^2\theta\tan^2\theta \ ,$$

to complete the proof.

████████████████ **Exercise 3I** ████████████████

1.   Use the TI-92 to draw the graphs of $\cos^2 x$, $\sin^2 x$ and $\sin^2 x + \cos^2 x$. Describe the relationships that exist between the three curves.

2.   (a)   Use the TI-92 to draw the graphs of $\cos x$ and $\sec x$ $(1/\cos x)$.

      (b)   Sketch graphs of $\sin x$ and $\csc x$ by hand. Then check your result with the TI-92.

      (c)   Repeat (ii) for $\tan x$ and $\cot x$.

3.   Solve the equations below, giving all solutions in the range $0° \leq \theta \leq 360°$.

      (a) $\sin^2 x + 3\cos^2 x = 2$ (e) $4\sin x - 2\csc x = \cot x$

      (b) $3\cos\theta - 2\sec\theta = 3\tan\theta$ (f) $5\sin^2 x + \sin x - 4 = 3\cos^2 x$

      (c) $\sin x + 4 = 7\cos^2 x$ (g) $2 + \cos^2 x = 3\sin x \cos x$

      (d) $6\sin^2 x - 4 = \cos x$ (h) $2\cot\theta = 3\tan\theta$

4.   Prove that

      $$1 + \cot^2\theta = \csc^2\theta$$

5.   Prove that

      $$\cos^2\theta - \sin^2\theta = 2\cos^2\theta - 1 \, ,$$

      hence show that

      $$\cos^4\theta - \sin^4\theta = 2\cos^2\theta - 1$$

6.   Prove each of the identities given below.

      (a) $\sin^2\theta - \cos^2\theta = 2\sin^2\theta - 1$ (d) $\cos\theta(\cot\theta + \tan\theta) = \csc\theta$

      (b) $(1-\cos\theta)(1+\cos\theta) = \sin^2\theta$ (e) $\sec^2\theta(\sec^2\theta - 1) = \tan^2\theta\,(1+\tan^2\theta)$

      (c) $(\sin\theta + \cos\theta)^2 + (\sin\theta - \cos\theta)^2 = 2$ (f) $\sin^2 A - \sin^2 B = \cos^2 B - \cos^2 A$

3.12   FURTHER IDENTITIES

There are a number of further identities that will not be proved in this book, but which are of importance in mathematics. The first group of these that we will consider are **Sum** and **Difference Formulae**, which are listed below.

$$\sin(A+B) = \sin A \cos B + \cos A \sin B$$

$$\sin(A-B) = \sin A \cos B - \cos A \sin B$$

$$\cos(A+B) = \cos A \cos B - \sin A \sin B$$

$$\cos(A-B) = \cos A \cos B + \sin A \sin B$$

$$\tan(A + B) = \frac{\tan A + \tan B}{1 - \tan A \tan B}$$

$$\tan(A - B) = \frac{\tan A - \tan B}{1 + \tan A \tan B}$$

It is a very simple matter to obtain a further set of identities known as the **Double Angle Formulae** from the previous set of identities, which give results for $\sin(2A)$, $\cos(2A)$ and $\tan(2A)$.

For $\sin(2A)$ the result is obtained as below:

$$\sin(2A) = \sin(A + A)$$
$$= \sin A \cos A + \cos A \sin A$$
$$= 2 \sin A \cos A .$$

For $\cos(2A)$:

$$\cos(2A) = \cos(A + A)$$
$$= \cos A \cos A - \sin A \sin A$$
$$= \cos^2 A - \sin^2 A .$$

By using the identity $\cos^2 A + \sin^2 A = 1$, this identity can also be written as

$$\cos(2A) = 2\cos^2 A - 1$$

or

$$\cos(2A) = 1 - 2\sin^2 A$$

TI-92 ACTIVITY 3E

(A) (i) Graph 2sin$x$ cos$x$. What are the amplitude and period of this wave? What sine wave would have this amplitude and period? Draw a graph of your answer to check.

(ii) Clear the GRAPH screen.
Graph $\cos^2 x - \sin^2 x$, $2\cos^2 x - 1$ and $1 - 2\sin^2 x$.
What are the amplitude and period of these waves?
What cosine wave would have this amplitude and period. Draw a graph of your answer to check.

(iii) Clear the GRAPH screen.
Graph $3\sin x - 4\sin^3 x$. What single sine wave is equivalent to the wave you have on the screen? Check your answer.

(B) (i) Clear the GRAPH screen.
Graph $\sin(5x) + \sin(3x)$. Graph $2\sin(4x) \cos(x)$. Comment on your results.

(ii) Repeat (i) for $\sin(9x) + \sin(3x)$ and $2\sin(6x) \cos(3x)$.

(iii) Express $\sin(7x) + \sin(3x)$ in the form $2\sin(Ax) \cos(Bx)$. Check your result by graphing $\sin(12x) - \sin(8x)$ and your prediction.

(iv) By plotting compare $\sin(10x) - \sin(4x)$ and $2\cos(7x) \sin(3x)$. Try to predict an alternative form for $\sin(12x) - \sin(8x)$. Check your result by graphing $\sin(12x) - \sin(8x)$ and your prediction.

The next set of identities are known as the **Factor Formulae** and are listed below.

$$\sin A + \sin B = 2\sin\left(\frac{A+B}{2}\right)\cos\left(\frac{A-B}{2}\right)$$

$$\sin A - \sin B = 2\cos\left(\frac{A+B}{2}\right)\sin\left(\frac{A-B}{2}\right)$$

$$\cos A + \cos B = 2\cos\left(\frac{A+B}{2}\right)\cos\left(\frac{A-B}{2}\right)$$

$$\cos A - \cos B = -2\sin\left(\frac{A+B}{2}\right)\sin\left(\frac{A-B}{2}\right)$$

▓▓▓▓▓▓▓▓▓▓ **Example 3J** ▓▓▓▓▓▓▓▓▓▓

For the triangles in Figure 3.28, show that

(i) $\cos(A+B) = \dfrac{33}{65}$

(ii) $\sin(A+B) = \dfrac{56}{65}$

(iii) $\tan(A+B) = \dfrac{56}{33}$

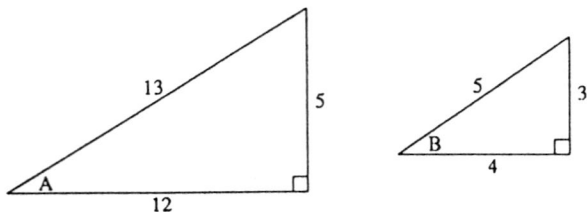

Figure 3.28

Solution

Using the sides of the first triangle

$$\cos A = \frac{12}{13}, \quad \sin A = \frac{5}{13}, \quad \tan A = \frac{5}{12},$$

and from the second triangle

$$\cos B = \frac{4}{5}, \quad \sin B = \frac{3}{5}, \quad \tan B = \frac{3}{4}.$$

(i)   Using the addition formula for $\cos(A+B)$

$$\cos(A+B) = \cos A \cos B - \sin A \sin B$$
$$= \frac{12}{13} \times \frac{4}{5} - \frac{5}{13} \times \frac{3}{5} = \frac{33}{65}.$$

(ii)   From the addition formula for $\sin(A+B)$

$$\sin(A+B) = \sin A \cos B + \cos A \sin B$$
$$= \frac{5}{13} \times \frac{4}{5} + \frac{12}{13} \times \frac{3}{5} = \frac{56}{63}.$$

(iii)   From the addition formula for $\tan(A+B)$

$$\tan(A+B) = \frac{\tan A + \tan B}{1 - \tan A \tan B}$$

$$= \frac{\dfrac{5}{12} + \dfrac{3}{4}}{1 - \dfrac{5}{12} \times \dfrac{3}{4}} = \frac{\dfrac{7}{6}}{\dfrac{33}{48}}$$

$$= \frac{56}{33}$$

▓▓▓▓▓▓▓▓▓▓ **Example 3K** ▓▓▓▓▓▓▓▓▓▓▓▓▓▓▓▓▓▓▓▓▓

Prove that

$$\sin(A+B)\sin(A-B) = \sin^2 A - \sin^2 B$$

Solution

Using the sum and difference formulae

$$\sin(A+B)\sin(A-B) = (\sin A \cos B + \cos A \sin B)(\sin A \cos B - \cos A \sin B)$$
$$= \sin^2 A \cos^2 B - \sin A \sin B \cos A \cos B$$
$$+ \sin A \sin B \cos A \cos B - \cos^2 A \sin^2 B$$
$$= \sin^2 A \cos^2 B - \cos^2 A \sin^2 B$$
$$= \sin^2 A(1 - \sin^2 B) - (1 - \sin^2 A)\sin^2 B$$
$$= \sin^2 A - \sin^2 A \sin^2 B - \sin^2 B + \sin^2 A \sin^2 B$$
$$= \sin^2 A - \sin^2 B$$

▓▓▓▓▓▓▓▓▓▓ **Example 3L** ▓▓▓▓▓▓▓▓▓▓▓▓▓▓▓▓▓▓▓▓▓

Prove that $\tan(x+45°) = \dfrac{1+\tan x}{1-\tan x}$

Solution

Using the appropriate addition formula gives

$$\tan(x+45°) = \frac{\tan x + \tan 45°}{1 - \tan x \tan 45°}$$

$$= \frac{\tan x + 1}{1 - \tan x} \qquad \text{(since } \tan 45° = 1\text{)}.$$

Example 3M

Prove that

$$\frac{\sin 10A - \sin 6A}{\cos 6A + \cos 5A} = \tan 2A$$

Solution

Applying the appropriate factor formula to the top and bottom of the LHS gives

$$\frac{\sin 10A - \sin 6A}{\cos 10A + \cos 6A} = \frac{2\cos\left(\dfrac{10A+6A}{2}\right)\sin\left(\dfrac{10A-6A}{2}\right)}{2\cos\left(\dfrac{10A+6A}{2}\right)\cos\left(\dfrac{10A-6A}{2}\right)}$$

$$= \frac{2\cos(8A)\sin(2A)}{2\cos(8A)\cos(2A)}$$

$$= \frac{\sin(2A)}{\cos(2A)}$$

$$= \tan(2A)$$

Exercise 3J

1.　For the triangle in Figure 3.29, find $\cos(A+B)$, $\sin(A+B)$ and $\tan(A+B)$.

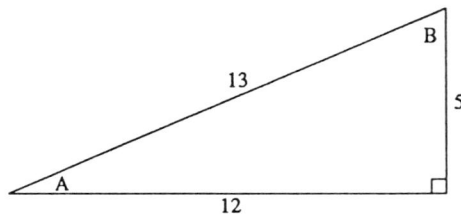

Figure 3.29

Comment on your results.

2.   If $\cos A = \dfrac{24}{25}$ and $\cos B = \dfrac{8}{17}$ find

    (a) $\sin A$ (e) $\sin(A{-}B)$ (i) $\cos(2A)$

    (b) $\sin B$ (f) $\tan(B{-}A)$ (j) $\sin(2B)$

    (c) $\tan A$ (g) $\cos(A{+}B)$

    (d) $\tan B$ (h) $\tan(A{+}B)$

3.   Repeat problem 2 given that $\cos A = 0.5$ and $\cos B = 0.8$.

4.   (a)   Show that $\sin(A + 90°) = \cos A$.

    (b)   Find simpler expressions for

        (i) $\sin(A + 180°)$ (iv) $\cos(A + 180°)$

        (ii) $\sin(A - 90°)$ (v) $\cos(A - 90°)$

        (iii) $\cos(A + 90°)$ (vi) $\tan(A - 45°)$

5.   Find alternative expressions for

    (a) $\cos 8x \sin x - \sin 8x \cos x$ (e) $\sin 40° \sin 10°$

    (b) $\sin 60° + \sin 20°$ (f) $\sin 69° \cos 40° - \sin 40° \cos 69°$

    (c) $\cos 50° - \cos 30°$ (g) $\sin 45° - \sin 35°$

    (d) $\sin 50° \cos 30°$ (h) $\sin 45° \cos x + \cos 45° \sin x$.

6.   (a)   Prove that $\sin(3A) = 3\sin A - 4\sin^3 A$.

    (b)   Find a similar result for $\cos(3A)$.

    (c)   Express $\sin(4A)$ in terms of $\sin A$ and $\cos A$.

7.   Prove that

(a) $\cos(A+B)\cos(A-B) = \cos^2 A - \sin^2 B$

(b) $\cos(A+B) + \cos(A-B) = 2\cos A \cos B$

(c) $\dfrac{\sin A}{\sin B} + \dfrac{\cos A}{\cos B} = \dfrac{2\sin(A+B)}{\sin(2B)}$

(d) $\cos(3\theta) - \cos(7\theta) = 2\sin(5\theta)\sin(2\theta)$

8.   Show that when two wave forms of roughly the same amplitude but slightly different periods are added together the result is a wave of roughly the same period but with an amplitude that varies with a frequency equal to the difference in frequencies of the two waves.

3.13   THE EXPRESSION $A\sin x + B\cos x$

TI-92 ACTIVITY 3F

(A)   (i)   Graph $\sin(x) + \cos(x)$. Using the ranges $-2\pi \le x \le 2\pi$ and $-2 \le y \le 2$.

(ii)   What is the amplitude of the wave that you produce? Call this value R.

(iii)   How far to the left of the y-axis does the curve cross the x-axis? Call this value α.

(iv)   Graph $R\sin(x+\alpha)$ using your values of R and α. What do you notice?

(B)   Repeat (A) for each combination given below.

(i) $2\sin(x) + 2\cos(x)$

(ii) $3\sin(x) + 4\cos(x)$

(iii) $5\sin(x) + 12\cos(x)$

(iv) $\sin(x) - \cos(x)$

(C)   Repeat (A) for each combination given below.

     (i) $3\sin(4x) + 4\cos(4x)$

     (ii) $5\cos(3x) + 12\sin(3x)$

     (iii) $\sin(2x) - 2\cos(2x)$

     (iv) $2\cos(5x) - 5\sin(5x)$

What do you deduce about the amplitude of the wave $A\sin(wx) + B\cos(wx)$?

Simplifying $A\sin x + B\cos x$

From the TI-92 Activity you will have seen that $A\sin x + B\cos x$ produces a sine wave of the form $R\sin(x+\alpha)$. Using the addition formulae it is possible to find relationships between the constants R and α and the constants A and B.

Using the addition formula

$$R\sin(x+\alpha) = R\sin x\cos\alpha + R\cos x\sin\alpha$$
$$= R\cos\alpha\sin x + R\sin\alpha\cos x.$$

But we know that

$$R\sin(x+\alpha) = A\sin x + B\cos x$$

Comparing the right hand sides of the above equations gives

$$R\cos\alpha = A \quad \text{and} \quad R\sin\alpha = B$$

To obtain an expression for R, begin by squaring and adding A and B, to give

$$A^2 + B^2 = R^2\cos^2\alpha + R^2\sin^2\alpha$$
$$= R^2(\cos^2\alpha + \sin^2\alpha)$$
$$= R^2$$

so

$$R = \sqrt{A^2 + B^2}$$

To obtain an expression for α, divide B by A to give

$$\frac{B}{A} = \frac{R\sin\alpha}{R\cos\alpha} = \frac{\sin\alpha}{\cos\alpha} = \tan\alpha$$

so

$$\alpha = \tan^{-1}\left(\frac{B}{A}\right)$$

Example 3N

(i)   Write $4\sin x + 3\cos x$ in the form $R\sin(x+\alpha)$.

(ii) Find the solutions of the equation

$$4\sin x + 3\cos x = 2.5\ ,$$

that lie in the range $0 \le x \le 360°$.

Solution

(i)   For $4\sin x + 3\cos x$ we have $A = 4$ and $B = 3$, so

$$R = \sqrt{4^2 + 3^2} = \sqrt{25} = 5 \quad \text{and} \quad \alpha = \tan^{-1}\left(\frac{3}{4}\right) = 36.9° \text{ or } 216.9°$$

To decide which value of α to use, note that $\cos\alpha = \frac{4}{5}$ and $\sin\alpha = \frac{3}{5}$. In this case both are positive so $0° < \alpha < 90°$. Hence α must take the value of $36.9°$.

(ii) The equation

$$4\sin x + 3\cos x = 2.5$$

can be rewritten as

$$5\sin(x + 36.9°) = 2.5$$

or

$$\sin(x + 36.9°) = 0.5$$

The solutions are then given by

$$x + 36.9° = 30°, 150°, 390°, \text{etc.}$$

so

$$x = -6.9°, 113.1°, 353.1°, \text{etc.}$$

The solutions in the required range are

$$x = 113.1° \quad \text{and} \quad 353.1°.$$

Exercise 3K

Solve the following problem 'by hand' and then check your answers using the TI-92.

1.   Express each of the following in the form $R\sin(x+\alpha)$.

(a) $12\sin x + 5\cos x$ (d) $6\sin x - 8\cos x$

(b) $12\sin x - 5\cos x$ (e) $5\sin x - \cos x$

(c) $\sin x + 2\cos x$ (f) $6\sin x - 3\cos x$

2.   By expanding $R\cos(x-\alpha)$, find an alternative form for $A\sin x + B\cos x$, expressing R and α in terms of A and B.

3.   Express each of the following in the form $R\cos(x-\alpha)$.

(a) $5\sin x + 12\cos x$ (c) $2\sin x - \cos x$

(b) $6\sin x - 3\cos x$ (d) $\sin x + 3\cos x$

4.   By referring back to problem 1, solve the equations below in the range $-180° \le x \le 180°$.

(a) $12\sin x + 5\cos x = 6$ (d) $6\sin x + 8\cos x = 10$

(b) $12\sin x - 5\cos x = 0$ (e) $5\sin x - \cos x = 1$

(c) $\sin x + 2\cos x = 1$ (f) $6\sin x - 3\cos x = 5$

4

Sequences and series

4.1 SEQUENCES

A sequence is defined as a set of numbers in a specified order, so that each number is related to the next by a rule. Each of the numbers in the sequence is called a **term**. Some examples of sequences are given below.

> 1, 8, 15, 22, 29, 36,....
> 1, 4, 9, 16, 25, 36,....
> $2, 1, \frac{1}{2}, \frac{1}{4}, \frac{1}{8}, \frac{1}{16}, \ldots$

An **expression** can be found for each of these sequences that describes how each term is calculated. The notation u_n is used to denote the nth term of the sequence. The first term is u_1, the second u_2, etc. Each of the sequences above is now given with a rule for u_n:

> 1, 8, 15, 22, 29, 36,.... $u_n = 7n - 6$
> 1, 4, 9, 16, 25, 36,.... $u_n = n^2$
> $2, 1, \frac{1}{2}, \frac{1}{4}, \frac{1}{8}, \frac{1}{16}, \ldots$ $u_n = 4(\frac{1}{2})^n$ or $u_n = (\frac{1}{2})^{n-2}$

The terms of the first two sequences given above increase indefinitely. The terms of the third sequence however get smaller and smaller, getting closer and closer to zero. We say that the limit of the sequence as n tends to infinity is zero. This is written as

$$\lim_{n \to \infty} \left(\tfrac{1}{2}\right)^{n-2} = 0$$

This means that as n tends to infinity (becomes larger and larger) the terms get closer and closer to zero, but never actually become zero.

TI-92 ACTIVITY 4A

It is possible to produce the terms of a sequence and find the limits of sequences using the TI-92. Clear the HOME screen and set MODE to EXACT.

(A)   The first activity introduces the use of the TI-92 in setting up and exploring the sequences. Consider the sequence $u_n = 4(\frac{1}{2})^n$.

   (i)   Type seq $(4(\frac{1}{2})^n, n, 1, 10)$ **ENTER**

          This will give the sequence $u_n = 4(\frac{1}{2})^n$ starting with $n = 1$ and ending with $n = 10$.

   (ii)   To obtain a particular term of the sequence, for example, u_5 type seq $(4(\frac{1}{2})^n, n, 5, 5)$.

   (iii) From the first ten terms of the sequence we might propose that the 'Limit' of the sequence is zero. This can be confirmed by using the Limit command.

          Type limit$(4(1/2)^\wedge n, n, \infty)$ **ENTER**

          (For limit press **F3** and select 3:limit.)

          Limit can also be used to find the nth term of the sequence, for example the 5th term, u_5, is obtained by

          limit$(4(1/2)^\wedge n, n, 5)$ **ENTER**

          Figure 4.1 shows the TI-92 screen for these first set of tasks.

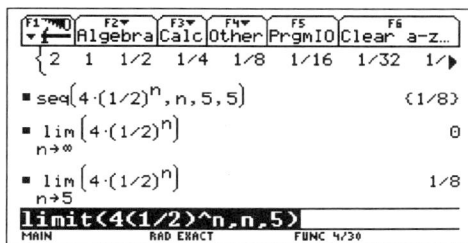

Figure 4.1   Exploring sequences on the TI-92

(B)   For each of the sequences defined below

(i)   Use the TI-92 to find the first 10 terms,

(ii)   write down the limit of the sequence if it exists,

(iii)   check your response to (ii) using the limit command on the TI-92.

(a) $u_n = \dfrac{1}{1+n}$ (b) $u_n = 4 + \dfrac{1}{n}$ (c) $u_n = \dfrac{n^3 + 1}{n}$

(d) $u_n = \dfrac{n^2 + 10}{n^2}$ (e) $u_n = 1.1^n$ (f) $u_n = 0.9^n$

Exercise 4A

1.   Below are listed four sequences of numbers.

(a)   4, 7, 10, 13, 16, ...

(b)   3, 6, 11, 18, 27, ...

(c)   0, 3, 8, 15, 24, ...

(d)   4, 9, 14, 19, 24, ...

Select the rule below which defines each sequence.

$$u_n = n^2 - 1 \quad u_n = n^2 + 2 \quad u_n = 5n - 1 \quad u_n = 3n + 1$$

2.   Write down the first five terms of each sequence defined below and state the
      limit of the sequence if one exists.

(a) $u_n = 2^n$ (b) $u_n = \left(\frac{1}{3}\right)^{n-1}$ (c) $u_n = (-2)^n$

(d) $u_n = \left(-\frac{1}{2}\right)^n$ (e) $u_n = \dfrac{2n+1}{n}$ (f) $u_n = n + (0.1)^n$

Check your answers on the TI-92.

3.   For each sequence find a rule for obtaining the nth term u_n. Also find the limit
      of each sequence if it exists.

(a)   1, 3, 9, 27, ...

(b)   14, 9, 4, −1, ...

(c)   5.1, 5.01, 5.001, 5.0001, ...

(d) $\frac{1}{3}, \frac{1}{2}, \frac{3}{5}, \frac{2}{3}, ...$

Check your answers on the TI-92.

4.2   SERIES

A **series** is formed when the terms of a sequence are added together. Some examples
of series are

    $1 + 4 + 9 + 16 + 25 + ...$
    $5 + 7 + 9 + 11 + 13 + ...$
    $1 + 2 + 4 + 8 + 16 + ...$

The symbol Σ is used to denote the sum of the terms of a series. The notation
$\sum_{i=1}^{n} u_i$ is used to denote the sum of the first n terms of the series. This can be expressed
as

$$\sum_{i=1}^{n} u_i = u_1 + u_2 + u_3 + \cdots + u_n$$

▓▓▓▓▓▓▓▓▓▓▓▓▓ **Example 4A** ▓▓▓▓▓▓▓▓▓▓▓▓▓▓▓▓▓▓▓▓▓▓▓▓▓▓▓▓

Write down the series described by

$$\sum_{i=1}^{6}(i^2-1)$$

and find the sum of this series.

Solution

The first six terms of the series must be evaluated.

$$u_1 = 1^2 - 1 = 0 \qquad\qquad u_4 = 4^2 - 1 = 15$$
$$u_2 = 2^2 - 1 = 3 \qquad\qquad u_5 = 5^2 - 1 = 24$$
$$u_3 = 3^2 - 1 = 8 \qquad\qquad u_6 = 6^2 - 1 = 35$$

Now these terms can be put into the series

$$\sum_{i-1}^{6}(i^2-1) = 0+3+8+15+24+35 = 85$$

So the sum of the series is 85.

▓▓

TI-92 ACTIVITY 4B

It is possible to find the sum of a series using the TI-92. The summation command
can be obtained using **F3** and selecting 4:Σ(sum.
Clear the HOME screen.

(A)   Press **F3**, select 4:Σ(sum and complete the expression Σ(3^i,i,1,7) **ENTER**.
       This command calculates the value of the summation

$$\sum_{i=1}^{7}3^i = 3+3^2+3^3+3^4+3^5+3^6+3^7 = 3279$$

(B)   Use the TI-92 to evaluate the following sums

(a) $\displaystyle\sum_{i=1}^{10}i$ (b) $\displaystyle\sum_{i=1}^{20}(i)^2$ (c) $\displaystyle\sum_{i=1}^{50}\frac{1}{i}$

(C)   Some series will converge to a limit when they are summed, but others will increase indefinitely or diverge. By using 4:Σ(sum with an upper limit of ∞ (inf), investigate the problems below.

(i)   Consider $\sum\limits_{i=1}^{\infty} \dfrac{1}{a^i}$ for different values of a. Try $a = 1, 2, 3$. For what values of a does the series converge to a limit?

(ii)   Consider $\sum\limits_{i=1}^{\infty} \dfrac{1}{i^n}$ for different values of n. Try $n = 1, 2, 3$. For what values of n does the series converge to a limit?

The convergence of infinite series is a fascinating subject which is not studied in this elementary text. Whether an infinite series will converge may not always be obvious from the sequence forming the series. For example, the series $\sum\limits_{i=1}^{\infty} \dfrac{1}{i}$ does not converge even though $\lim\limits_{n\to\infty} u_n = 0$ (where $u_n = \dfrac{1}{n}$). There are many mathematical tests for convergence. A simple test is that the series $\sum\limits_{i=1}^{\infty} u_i$ will converge if $\lim\limits_{n\to\infty} \left| \dfrac{u_{n+1}}{u_n} \right| < 1$. This is called the **ratio test**. Note that for $\sum\limits_{i=1}^{\infty} \dfrac{1}{i}$, we have $\lim\limits_{n\to\infty} \dfrac{u_{n+1}}{u_n} = 1$.

Exercise 4B

1.   Write each of the series below in full and find their sum.

(a) $\sum\limits_{i=1}^{4} (i+1)^2$ (b) $\sum\limits_{i=1}^{6} \dfrac{1}{i}$ (c) $\sum\limits_{i=1}^{4} (4i-2)$

(d) $\sum\limits_{i=1}^{5} 2i$ (e) $\sum\limits_{i=1}^{3} \dfrac{1}{i^2}$ (f) $\sum\limits_{i=1}^{4} (i^3 - 3)$

Check your answers on the TI-92.

2.   Write each of the series given below using sigma notation.

(a) $16 + 11 + 6 + 1 - 4$ (b) $2 + 6 + 18 + 54$

(c) $\tfrac{1}{2} + 2 + 4\tfrac{1}{2} + 8 + 12\tfrac{1}{2} + 18$ (d) $1 - 2 + 4 - 8 + 16$

3.   Apply the ratio test to each of the series in problem 1 to suggest which infinite series would converge to a limit. For those which do converge use the TI-92 to find the limit.

4.3   ARITHMETIC PROGRESSIONS

An **arithmetic progression** (AP) is a sequence where each term is obtained from the previous term by adding or subtracting the same number each time. The example below is a sequence where 3 is added each time to obtain the next term.

2, 5, 8, 11, 14, 17

The number that is added each time is known as the **common difference** and usually represented by the letter d. The first term is usually represented by the letter a.

In general an arithmetic progression will have terms given by

$a, a + d, a + 2d, a + 3d, ...$

So the nth term u_n will be given by

$u_n = a + (n-1)d$

Example 4B

An AP has first term 5 and common difference 4.

(a)   Write down the first 6 terms of the AP.

(b)   Find an expression for u_n, the nth term.

(c)   Find the 50th term using the answer to (b).

Solution

(a)   The AP will begin with 5 and subsequent terms can be obtained by adding 4 to the previous term to give

5, 9, 13, 17, 21, 25

(b)   The nth term is given by

$$u_n = a + (n-1)d$$

Here $a = 5$ and $d = 4$ so,

$$u_n = 5 + (n-1)4 = 5 + 4n - 4 = 1 + 4n$$

(c)   The 50th term is given by u_{50}.

$$u_{50} = 1 + 4 \times 50 = 201$$

Example 4C

The sequence

2, 4, 6, 8, 10, 12, 14, 16, 18, 20

is an AP. Find the sum of its terms.

Solution

The sum of its terms, S, is given by the series

$$S = 2 + 4 + 6 + ... + 16 + 18 + 20$$

Consider the series written in reverse order.

$$S = 20 + 18 + 16 + ... + 6 + 4 + 2$$

Now adding the two series together gives

$$2S = 22 + 22 + 22 + ... + 22 + 22 + 22$$

Noting that the series has 10 terms gives

$$2S = 10 \times 22$$

so

$$S = \frac{10 \times 22}{2} = 110.$$

The Sum of an Arithmetic Progression

Using the approach of Example 4C we can find a general formula for the sum of an AP. The sum of the first n terms of an AP, S_n, is given by

$$S_n = a + (a+d) + \cdots + \big(a + (n-2)d\big) + \big(a + (n-1)d\big)$$

Writing the series in reverse order

$$S_n = \big(a + (n-1)d\big) + \big(a + (n-2)d\big) + \cdots + (a+d) + a$$

and adding the two series gives

$$2S_n = \big(a + a + (n-1)d\big) + \big(a + d + a + (n-2)d\big) + \cdots + \big(a + d + a + (n-2)d\big) + \big(a + a + (n-1)\big)$$

$$2S_n = \big(2a + (n-1)d\big) + \big(2a + (n-1)d\big) + \cdots + \big(2a + (n-1)d\big) + \big(2a + (n-1)d\big)$$

There are n identical terms on the right hand side so that

$$S_n = \frac{n\big(2a + (n-1)d\big)}{2}$$

The sum of an AP

$$u_n = a + (n-1)d$$

is

$$S_n = \frac{n\big(2a + (n-1)d\big)}{2}$$

░░░░░░░░░░░░░░░░░ **Example 4D** ░░░░░░░░░░░░░░░░░░░░░░░░░░░░░░

Find the sum of the arithmetic progression

$$1 + 8 + 15 + ... + 260$$

Solution

This is an AP with $a = 1$ and $d = 7$. To find n use

$$u_n = a + (n - 1)d$$

applied to the last term. This gives

$$260 = 1 + 7(n-1)$$

$$260 = 7n - 6$$

$$n = \frac{260 + 6}{7}$$

$$= 38.$$

Now the sum can be found

$$S_{38} = \frac{38\left(2 \times 1 + (38 - 1) \times 7\right)}{2}$$

$$= 4959$$

░░░░░░░░░░░░░░░░░ **Exercise 4C** ░░░░░░░░░░░░░░░░░░░░░░░░░░░░░

1.   Find the 8th term of each of the APs given below.

   (a)   1, 10, 19, 28,..

   (b)   4, 3, 2, 1,..

   (c)   5, 7, 9, 11,..

   Also write down an expression for the nth term of each AP.

2.   For each AP below find the common difference and the number of terms.

   (a)   4, 7, 10, ..., 85

   (b)   −1, 3, 7, ..., 67

   (c)   51, 44, 37, ..., −54

3.   Find the sum of each of the following series.

   (a) $1 + 4 + 7 + 10 + ... + 70$

   (b) $6 + 4 + 2 + ... - 18$

   (c) $10 + 10.1 + 10.2 + ... + 20$

4.   Find the sum of the first 10 terms of the APs that begin

   (a)   1, 3, 5, ...

   (b)   7, 10, 13, ...

5.   Find the sum of the first 20 even numbers.

6.   In an AP the fifth term is 16 and the second term 7. Find the first term.

7.   The fifth term of an AP is 22 and the sum of the first 10 terms is 240. Find the first term and the common difference.

4.4   GEOMETRIC PROGRESSIONS

In a **geometric progression** (GP) each term is obtained from the previous term by multiplying by the same number each time. For example the sequence below is obtained by multiplying the previous term by 2.

   3, 6, 12, 24, 48, 96, ...

The number that is used to multiply the terms of the GP is known as the **common ratio**, usually referred to as r. In the above example the common ratio is 2. A GP with first term a and common ratio r would be

   $a, ar, ar^2, ar^3, ar^4, ...$

The nth term of such a sequence would be given by $u_n = ar^{n-1}$.

Example 4E

The first two terms of a GP are 16 and 22.4.

(a)   Find the common ratio.

(b)   Give an expression for the nth term.

(c)   Find the 4th term of the GP.

Solution

(a)   To find the common ratio note that the first term is a and the second ar, so

$$r = \frac{ar}{a} = \frac{u_2}{u_1}$$

In fact the ratio of any 2 successive terms could be used. In this case

$$r = \frac{22.4}{16} = 1.4$$

(b)   The nth term is given by

$$u_n = ar^{n-1}$$

so here

$$u_n = 16 \times 1.4^{n-1}$$

(c)   The 4th term is given by

$$u_4 = 16 \times 1.4^3$$
$$= 16 \times 2.744$$
$$= 43.904 .$$

The Sum of a GP

We now look for a general formula for the sum of a GP with n terms. Using S_n to denote this sum,

$$S_n = a + ar + ar^2 + ... + ar^{n-1}$$

This expression is now multiplied by r to give

$$rS_n = ar + ar^2 + ar^3 + ... + ar^{n-1} + ar^n$$

Subtracting the second of these expressions from the first gives

$$S_n - rS_n = a - ar^n$$

or

$$S_n(1 - r) = a(1 - r^n)$$

so that

$$S_n = \frac{a(1-r^n)}{(1-r)}$$

Example 4F

Find the sum of the first 20 terms of the GP

4, 6.8, 11.56, 19.652, ...

Solution

Here $a = 4$. The common ratio r is given by

$$r = \frac{6.8}{4} = 1.7$$

So the sum of the first 20 terms, S_{20}, is given by

$$S_{20} = 4\frac{(1-1.7^{20})}{(1-1.7)} = 232236 \qquad \text{to 6 significant figures.}$$

The Sum of an Infinite Geometric Progression

To find the sum of a GP with an infinite number of terms we consider the limit of S_n as $n \to \infty$.

$$\lim_{n \to \infty} a\left(\frac{1-r^n}{1-r}\right)$$

As only the r^n part of this expression depends on n, it is the behaviour of this that is important. If $-1 < r < 1$ then as $n \to \infty$, $r^n \to 0$, so

$$\lim_{n \to \infty} a\left(\frac{1-r^n}{1-r}\right) = a\left(\frac{1-0}{1-r}\right)$$

$$= \frac{a}{1-r}.$$

If $r > 1$ then $r^n \to \infty$ as $n \to \infty$.

If $r < -1$ then r^n will alternate between positive and negative values, however $|r^n|$ will also become bigger and bigger.

So the sum of an infinite GP will only converge if $-1 < r < 1$.

Example 4G

The first four terms of an infinite GP are given by

$$100, 99, 98.01, 97.0299$$

Find its sum.

Solution

Here $a = 100$ and $r = 0.99$.

The sum S is given by

$$S = \frac{a}{1-r} = \frac{100}{1-0.99} = \frac{100}{0.01} = 10000.$$

Exercise 4D

1.   For each of the geometric progressions below, find

      (i)   the common ratio,
      (ii)   the 10th term,
      (iii)   the sum of the first 8 terms.

      (a)   4, 4.8, 5.76, ...
      (b)   1.3, 1.04, 0.832, ...
      (c)   −1.7, 0.85, −0.425, ...
      (d)   200, 320, 512, ...

2.   For each GP below find the sum to infinity, if it exists.

      (a)   40, 16, 6.4, ...

      (b)   0.5, 0.6, 0.72, ...

      (c)   1, 0.9, 0.81, 0.729

      (d)   −1, 0.9, −0.81, 0.729, ...

3.   A ball is dropped from a height of 1m. Each time it bounces it rebounds to 2/5 of the height from which it was released. How far does it travel before it stops?

4.   The second and third terms of a geometric progression are 12 and $6(c+1)$ respectively. Find c if the sum of the first three terms of the GP is 38.

5.   Show that if you invest £1000 in a building society at an annual interest rate of r% then the amount after n years forms the geometric progression

$$u_n = 1000\left(1 + \frac{r}{100}\right)^n$$

[Assume that you do not remove any money!].

4.5   THE BINOMIAL THEOREM

In this section we deal with expanding expressions like $(2+x)^{12}$ or $(5-x)^4$. This is then developed to enable series expansions to be made for expressions like $(1-x)^{\frac{1}{2}}$ or $(1+x^2)^{-2}$.

Expanding $(a+b)^n$: n a Positive Integer

TI-92 ACTIVITY 4C

The aims of this Activity are to investigate the coefficients in the expansion of $(a + b)^n$. Clear the HOME screen and select AUTO mode.

(A)   (i)   With the TI-92 use **F2** 3:expand(to find the expansion of $(1 + x)^2$. Expand the following expressions:

$$(1 + x)^3, (1 + x)^4, (1 + x)^5.$$

What do you notice about the powers of x that you obtain?
What do you notice about the coefficients of the terms you obtain?

(ii)   If you expand $(1 + x)^6$ what would you expect the first 2 and the last 2 terms to be?

Try it to check your prediction.

(iii)   The number pattern below is known as **Pascal's Triangle**.

$$
\begin{array}{ccccccccccc}
 & & & & & 1 & & 1 & & & \\
 & & & & 1 & & 2 & & 1 & & \\
 & & & 1 & & 3 & & 3 & & 1 & \\
 & & 1 & & 4 & & 6 & & 4 & & 1 \\
 & 1 & & 5 & & 10 & & 10 & & 5 & & 1
\end{array}
$$

How do the coefficients of your expressions relate to the triangle?
What would you expect the next two rows of the triangle to be?
Now expand $(1 + x)^7$ to check your prediction.

(B)   (i)   Expand $(a+b)^1$, $(a+b)^2$, $(a+b)^3$, $(a+b)^4$.
Do the coefficients still relate to Pascal's Triangle?
What happens to the powers of a and b?

(ii)   Using Pascal's Triangle write down the expansion of $(a+b)^5$. Check your prediction on the TI-92.

(C)   (i)   Pascal's Triangle provides a good approach for small values of n. Now consider $(a + b)^{12}$. The TI-92 can expand this easily, try it and see. We need an easy method of finding the coefficients in the expansion of $(a + b)^n$.

(ii)   Consider the expansion of $(a + b)^4 = (a + b)(a + b)(a + b)(a + b)$

$$a^4 + 4a^3b + 6a^2b^2 + 4ab^3 + b^4$$

The coefficient of a^4 is the number of ways we can form a^4 from the brackets, there is only one.

The coefficient of a^3b is the number of ways we can choose a from 3 brackets and b from one bracket. There are four ways: *aaab, aaba, abaa, baaa.*

The coefficient of a^2b^2 is the number of ways we can choose a from 2 brackets and b from the other two brackets. There are six ways: *aabb, abab, baab, abba, baba, bbaa.*

There is a convenient method for calculating the number of ways of choosing objects in this way.

Notation

The coefficient of $a^{n-r}b^r$ in the expansion of $(a + b)^n$ is the number of ways of choosing b from r brackets and a from $(n - r)$ brackets, and is calculated by

$$\frac{n!}{r!(n-r)!}$$

and denoted by ^nC_r.

$n!$ is called 'n factorial' and is the product $n(n - 1)(n - 2) \dots 3.2.1$
e.g. $6! = 6 \times 5 \times 4 \times 3 \times 2 \times 1 = 720$.

(iii)   Use the definition of ^nC_r to calculate $^4C_1$, $^4C_2$, $^4C_3$ and $^4C_4$. (Note that $0! = 1$.) Are these values the coefficients in $(a + b)^4$?

(iv)   Use the definition of ^nC_r to calculate $^5C_1$, $^5C_2$, $^5C_3$, $^5C_4$ and $^5C_5$. Are these values the coefficients in $(a + b)^5$?

(v)   The TI-92 evaluates ^nC_r directly.
        Type $ncr(5,2)$ **ENTER**.
        Which coefficient in $(a + b)^5$ does it give?
        Use the TI-92 and the $ncr(\ )$ command to find the coefficients in $(a + b)^7$.

(vi)   Type seq($ncr(7,i),i,0,7$) **ENTER**.
        What does the sequence mean?

(vii)   Use the TI-92 to evaluate $ncr(n,0)$, $ncr(n,1)$, $ncr(n,2)$, $ncr(n,3)$.
        What do you notice?
        Predict the formula for $ncr(n,r)$ for any integer r.

(viii) The factorial symbol ! can be obtained using

                MATH (2nd 5) select 7:Probability and 1:!

        Evaluate 3!, 6!, 10!

Summary

The expansion of $(a+b)^n$ where n is a positive integer is given by

$$(a+b)^n = {^nC_0}a^n + {^nC_1}a^{n-1}b + {^nC_2}a^{n-2}b^2 + \cdots + {^nC_n}b^n$$

where $^nC_r = \dfrac{n!}{r!(n-r)!}$. This is called **the Binomial Theorem**.

Example 4H

Expand each of the expressions below using the **Binomial Theorem**.

(a) $(1+x)^6$

(b) $(2+3x)^4$

Solution

(a)   Taking $a = 1$ and $b = x$ gives

$$(1+x)^6 = {}^6C_0 1^6 x^0 + {}^6C_1 1^5 x^1 + {}^6C_2 1^4 x^2 + {}^6C_3 1^3 x^3$$
$$+ {}^6C_4 1^2 x^4 + {}^6C_5 1^1 x^5 + {}^6C_6 1^0 x^6$$
$$= 1 + 6x + 15x^2 + 20x^3 + 15x^4 + 6x^5 + x^6$$

(b)   Taking $a = 2$ and $b = 3x$ gives

$$(2+3x)^4 = {}^4C_0 2^4 (3x)^0 + {}^4C_1 2^3 (3x)^1 + {}^4C_2 2^2 (3x)^2 + {}^4C_3 2^1 (3x)^3$$
$$+ {}^4C_4 2^0 (3x)^4$$
$$= 1 \times 16 \times 1 + 4 \times 8 \times 3x + 6 \times 4 \times 9x^2 + 4 \times 2 \times 27x^3 + 1 \times 1 \times 81x^4$$
$$= 16 + 96x + 216x^2 + 216x^3 + 81x^4 .$$

Example 4I

Find the first four terms of the expansion of $(1+x)^n$.

Solution

Note first that

$$^nC_0 = \frac{n!}{0!(n-0)!} = 1$$

$$^nC_1 = \frac{n!}{1!(n-1)!} = n$$

$$^nC_2 = \frac{n!}{2!(n-2)!} = \frac{n(n-1)}{2!}$$

$$^nC_3 = \frac{n!}{3!(n-3)!} = \frac{n(n-1)(n-2)}{3!}$$

The first four terms will be given by

$$(1+x)^n = {}^nC_0 + {}^nC_1 x + {}^nC_2 x^2 + {}^nC_3 x^3$$
$$= 1 + nx + \frac{n(n-1)}{2!} x^2 + \frac{n(n-1)(n-2)}{3!} x^3$$

Note that the pattern of terms will continue with the rth term given by

$$\frac{n(n-1)(n-2)\cdots(n-r+1)}{r!}x^r$$

Also note that as n is a positive integer, then the series will terminate when $n = r$.

▓▓▓▓▓▓▓▓▓▓▓ **Exercise 4E** ▓▓▓▓▓▓▓▓▓▓▓▓▓▓▓▓▓▓▓▓▓▓▓▓▓▓▓

Solve each of the following problems 'by hand' and then check your answers using the TI-92.

1.   Expand each of the following.

   (a) $(1+x)^4$ (b) $(2+x)^3$

   (c) $(1-x)^5$ (d) $(5+3x)^4$

   (e) $(5-2x)^3$ (f) $(x-y)^5$

   (g) $\left(u+\dfrac{1}{u}\right)^4$ (h) $\left(2x+\dfrac{1}{x}\right)^4$

2.   Find the coefficient of x^3 in each of the expansions below.

   (a) $(2+5x)^{10}$ (b) $(3-2x)^4$

   (c) $(8-7x)^3$ (d) $(3-2x^3)^2$

3.   Find the coefficient of x^7 in each of the expansions below.

   (a) $(1+x)^{13}$ (b) $(1-2x)^{11}$

   (c) $(3+0.5x)^{13}$ (d) $(1.3-0.1x)^9$

   (e) $(a+bx)^{15}$ (f) $(a-bx)^{14}$

4.   Write each of the following in the form $(a + bx)^n$.

(a) $1 + 6x + 15x^2 + 20x^3 + 15x^4 + 6x^5 + x^6$

(b) $0.0625 + 0.5x + 1.5x^2 + 2x^3 + x^4$

(c) $1 - 2x + 1.5x^2 - 0.5x^3 + 0.0625x^4$

The Binomial Theorem for Any Index

TI-92 ACTIVITY 4D

The previous section has shown how the Binomial Theorem can be applied to expand $(1+x)^n$, where n is a positive integer. Now we will investigate what happens for other values of n. Clear the HOME screen.

(A)   Define (**F4** select 1:Define) the function $f(n,x)$ by

$$f(n,x) = 1 + nx + \frac{n(n-1)x^2}{2} + \frac{n(n-1)(n-2)x^3}{6}$$

Evaluate $f(\tfrac{1}{2},x)$ and Define $y1(x) = f(\tfrac{1}{2},x)$.
Your HOME screen should look like Figure 4.2.

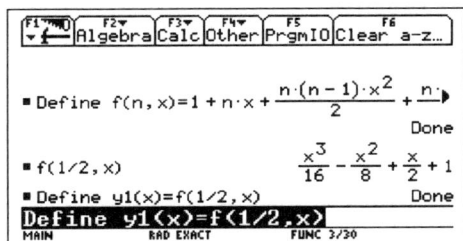
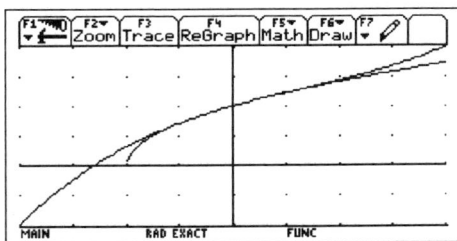

Figure 4.2   Figure 4.3

Comparing $f(\tfrac{1}{2},x)$ with $(1+x)^{\frac{1}{2}}$

In the Y = Editor define $y2(x) = (1+x)^{\frac{1}{2}}$.
Choose the WINDOW

xmin = −2, xmax = 2, xscl = .5
ymin = −1, ymax = 2, yscl = .5

Graph the functions $y1(x)$ and $y2(x)$.
Your GRAPH screen should look like Figure 4.3.

Use Zoom In (**F2** 2:Zoom In) to find the range of values of x for the series $f(n,x)$ to give a good approximation to the function $(1+x)^{\frac{1}{2}}$.

(B)   Extend your binomial series by adding the terms

$$\frac{n(n-1)(n-2)(n-3)x^4}{4!}+\frac{n(n-1)(n-2)(n-3)(n-4)x^5}{5!}$$

Investigate how this improves the quality of the approximation.
You may wish to add further terms.

(C)   Substitute each value of n given below into the expression you authored in (A) (i). Graph it and then compare it with the graph of the function $(1+x)^n$. Comment on the range of values for which the series gives a good approximation.

(i) $n=-1$ (ii) $n=\frac{3}{2}$

(iii) $n=\frac{1}{3}$ (iv) $n=-2$

Can you see a pattern emerging? What can you deduce about the binomial theorem for negative and fractional powers of n?

Summary

You have seen that $(1+x)^n$ gives a finite series when n is a positive integer. The notation ^nC_r would have no meaning when n is not a positive integer, but it is possible to produce a series expansion of $(1+x)^n$ for other values of n by using

$$(1+x)^n = 1+nx+\frac{n(n-1)x^2}{2!}+\frac{n(n-1)(n-2)}{3!}x^3+\cdots$$

Here neither $n!$ or ^nC_r are used. You will have seen in TI-92 Activity 4D that this type of series gives a good approximation to $(1+x)^n$ for small values of x.

It is important to note that for fractional or negative values of n the series will never terminate and will have an infinite number of terms. The series will only converge if $-1 < x < 1$ or $|x| < 1$. The more terms that you take in a series, the better approximation it will give to $(1+x)^n$.

Example 4J

Find the first four terms of the expansion of $(1+x)^n$ for (a) $n = -1$ and (b) $n = \frac{1}{2}$.

Solution

The terms required are given below

$$(1+x)^n = 1 + nx + \frac{n(n-1)x^2}{2!} + \frac{n(n-1)(n-2)}{3!}x^3 + \cdots$$

(a)   For $n = -1$

$$(1+x)^n = 1 + (-1)x + \frac{(-1)(-1-1)x^2}{2!} + \frac{(-1)(-1-1)(-1-2)x^3}{3!} + \cdots$$

$$= 1 - x + x^2 - x^3 + \cdots$$

(b)   For $n = \frac{1}{2}$

$$(1+x)^{\frac{1}{2}} = 1 + (\tfrac{1}{2})x + \frac{(\tfrac{1}{2})(\tfrac{1}{2}-1)x^2}{2!} + \frac{(\tfrac{1}{2})(-\tfrac{1}{2}-1)(-\tfrac{1}{2}-2)x^3}{3!} + \cdots$$

$$= 1 + \frac{x}{2} - \frac{x^2}{8} + \frac{x^3}{16} + \cdots$$

Example 4K

Expand $(1+3x)^{-2}$ using the binomial theorem and state the range of values for which it converges.

Solution

Using the expansion,

$$(1+x)^n = 1 + nx + \frac{n(n-1)x^2}{2!} + \cdots$$

with $n = -2$ gives

$$(1+3x)^{-2} = 1 + (-2)(3x) + \frac{(-2)(-2-1)(3x)^2}{2!} + \frac{(-2)(-2-1)(-2-2)(3x)^3}{3!}$$
$$= 1 - 6x + 27x^2 - 108x^3 + \cdots$$

For the series to converge $-1 < 3x < 1$ or $-\frac{1}{3} < x < \frac{1}{3}$.

Exercise 4F

1. Find the first four terms of the binomial expansion for each expression.

   (a) $(1+x)^{-2}$ (b) $(1+x)^{-3}$

   (c) $(1+x)^{3/2}$ (d) $(1+x)^{-1/2}$

2. State the values of x for which a binomial expansion of the following expressions would converge.

   (a) $\dfrac{1}{1-2x}$ (b) $\sqrt{1+\frac{1}{3}x}$

   (c) $\dfrac{1}{(1+4x)^2}$ (d) $\dfrac{1}{\sqrt{1-5x}}$

3. Find the first four terms of the binomial expansions of each expression, stating the range of values of x for which it converges.

   (a) $(1+3x)^{\frac{1}{2}}$ (b) $(1-2x)^{-2}$

   (c) $\dfrac{1}{1+4x}$ (d) $\dfrac{1}{\sqrt{1+2x}}$

4. Use the expansion of $(1+x)^{\frac{1}{2}}$ obtained in Example 4J to evaluate $\sqrt{0.98}$ and $\sqrt{1.2}$ by substituting suitable values of x.

5.   (a)   Show that

$$\frac{1}{(3+x)^2} = \frac{1}{x^2\left(\dfrac{3}{x}+1\right)^2}$$

    (b)   Find the first four non–zero terms of the expansion of $\left(\dfrac{3}{x}+1\right)^{-2}$.

    (c)   State the range of values of x for which this expansion converges.

    (d)   Use your answers to find a series expansion for $(3+x)^{-2}$.

6.   Find expansions for each of the expressions below.

    (a) $\dfrac{1}{2+x}$ (b) $\dfrac{1}{4+2x}$

    (c) $\dfrac{1}{(5+x)^2}$ (d) $\sqrt{2+x}$

5

Simple numerical methods for solving equations

5.1   DECIMAL SEARCH

TI-92 ACTIVITY 5A

The aim of this Activity is to show how the calculator can be used to search for the solution of an equation. Clear the HOME screen and Y = Editor. Select AUTO mode.

(A)   Suppose that you have to find the solutions (or roots) of the equation $3x^3 + x^2 - 5x - 1 = 0$.

  (i)   On the TI-92 set up and Graph $y1(x) = 3x^3 + x^2 - 5x - 1$ as shown in Figure 5.1. From the graph it should be clear that one solution lies between -2 and -1, another between -1 and 0 and the third solution between 1 and 2.

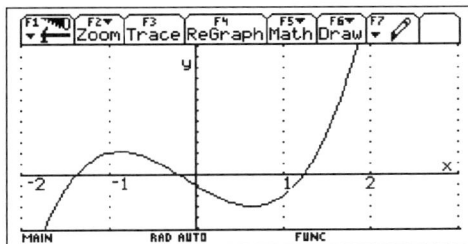

Figure 5.1   Graph of $y1(x) = 3x^3 + x^2 - 5x - 1$

  (ii)   We will concentrate on finding the positive solution. Return to the HOME screen and evaluate $y1(1)$ and $y1(2)$. Note that one gives a positive value and the other a negative value. This shows that the curve must cross the axis between these two points.

(iii)   Now evaluate $y1(1.2)$ and $y1(1.3)$.   What can you say about the solution?

(iv)   Now evaluate $y(1.23)$ and $y1(1.24)$.   What can you conclude about the solution?

(v)   Now try 1.234, 1.235 and 1.2345.

(vi)   Give the solution correct to 3 decimal places.

(vii)   By trying further values of your own find the solution correct to 4 decimal places.

(B)   Use an approach similar to (A), but substituting values of your own choice to find the other two roots of the equation.

(C)   Use an approach similar to that described in (A) to find the solutions of

(i) $x^3 + 3x^2 - 1$ (positive solution only)

(ii) $e^x - x^2 = 0$

(iii) $x - \cos x = 0$.

(D)   This decimal search method of approach can be used directly with the TI-92's TABLE facility.
Press ♦ **Y** (i.e. choose TABLE).
Press **F2** and choose starting value 1 and steplength (Δtbl) equal to 0.1 as shown in Figure 5.2. Press **ENTER**.

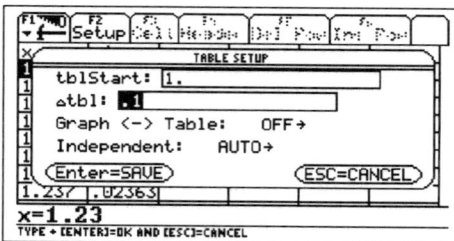

Figure 5.2
Setting up the starting value
and step length

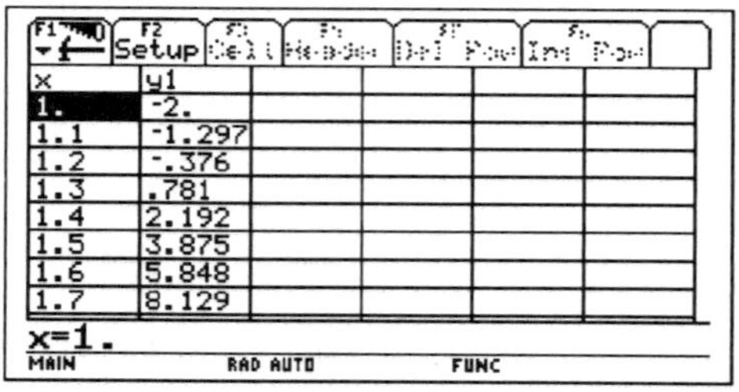

Figure 5.3
Table of values for $y1(x)$

Figure 5.3 shows the table of values for $y1(x)$ starting at $x = 1$. We see that the function values change sign between $x = 1.2$ and $x = 1.3$. The solution is between these values.

Now change the starting value to 1.2 and step to 0.01. Figure 5.4 shows the new table of values. It shows that the solution lies between 1.23 and 1.24. Change the starting value to 1.23 and step to 0.001. The table in Figure 5.5 shows the solution is between 1.234 and 1.235. The process can be repeated to find the solution to a chosen accuracy.

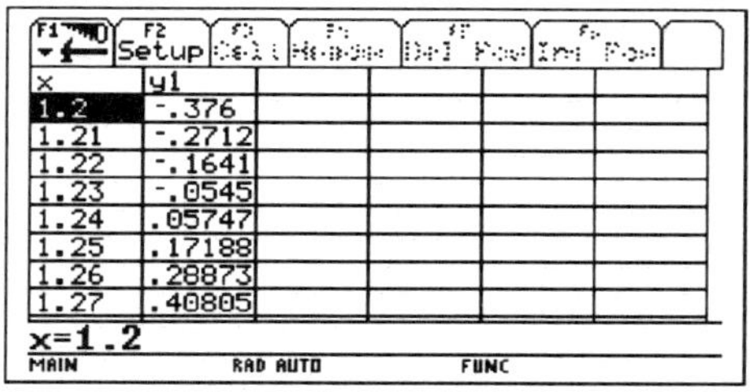

x	y1
1.2	-.376
1.21	-.2712
1.22	-.1641
1.23	-.0545
1.24	.05747
1.25	.17188
1.26	.28873
1.27	.40805

x=1.2
MAIN   RAD AUTO   FUNC

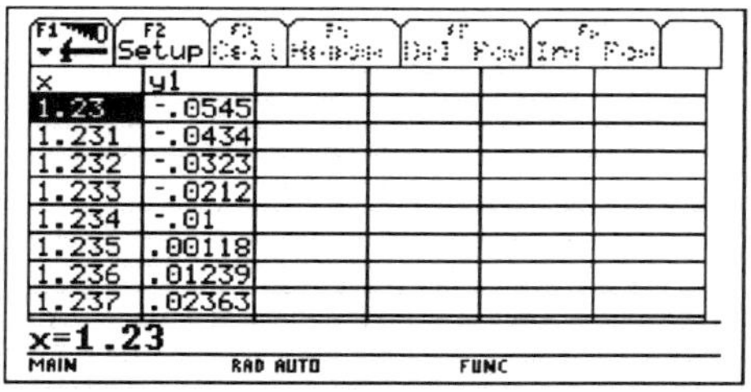

x	y1
1.23	-.0545
1.231	-.0434
1.232	-.0323
1.233	-.0212
1.234	-.01
1.235	.00118
1.236	.01239
1.237	.02363

x=1.23
MAIN   RAD AUTO   FUNC

Figure 5.4   Figure 5.5
Starting at 1.2   Starting at 1.23

(E)   Use the TABLE facility on the TI-92 to find the solutions in part (C).

Summary

The TI-92 activity has shown you how to find the solutions of the equation $f(x) = 0$, by trying a range of values of x. If $f(x)$ changes from being positive to negative between any two values, then a solution must lie between these two values. Further intermediate values of x can then be tried until the solution is obtained to the required degree of accuracy. This is called the **decimal search** method.

Example 5A

For each of the functions f find the solution of $f(x) = 0$ that lies between 1 and 2.

(a) $f(x) = \dfrac{x}{2} - \dfrac{1}{x}$

(b) $f(x) = x^3 - 7$

Solution

(a)   Table 5.1 shows the process for $f(x) = \dfrac{x}{2} - \dfrac{1}{x}$.

From $f(1) < 0$ and $f(2) > 0$ we deduce that the solution lies between $x = 1$ and $x = 2$.

Since $f(1.5) > 0$ we deduce that the solution lies between $x = 1$ (for which $f(1) < 0$) and $x = 1.5$.
And so on.
The third column (comments on solution) summarises this information.

Table 5.1

Trial Value of x	Value of $f(x)$ (to 4 dp)	Comments on solution
1	−0.5	−
2	0.5	$1 < x < 2$
1.5	0.0833	$1 < x < 1.5$
1.4	−0.0143	$1.4 < x < 1.5$
1.45	0.0353	$1.4 < x < 1.45$
1.42	0.0058	$1.4 < x < 1.42$
1.41	−0.0042	$1.41 < x < 1.42$
1.415	0.0008	$1.41 < x < 1.415$

The solution can now be quoted as 1.41 correct to 2 decimal places. Further work would lead to a more accurate solution.

(b)   Figure 5.6 shows the approach using the TI-92's TABLE facility. This has starting value 1.912 and step 0.0001. We see that the solution lies between 1.9129 and 1.9130.

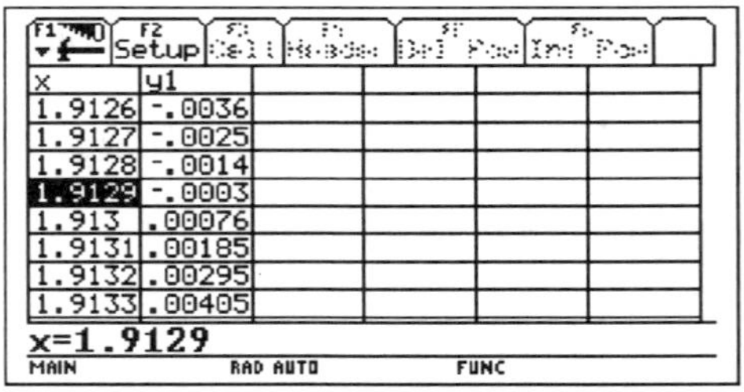

Figure 5.6   Solving $x^3 - 7 = 0$ using TABLE

Now the solution can be given as 1.913 correct to 3 decimal places.

1. Show that the equation $x^4 - 5$ has a solution between 1 and 2. Find this solution correct to 2 decimal places.

2. Show that the equation $x^3 - x + 10 = 0$ has a solution between -2 and -3. Find this solution correct to 2 decimal places.

3. Show that the equation $\sin(x) - x + 1 = 0$ has a solution between 1 and 2. Find this solution correct to 3 decimal places working in radians.

4. Sketch a graph of $y = e^x + x$. Use this to help you find the solution of $e^x + x = 0$.

5. (a) Find the solution of the equation $x^3 - 10 = 0$.

   (b) Find the cube root of 20.

5.2   FORMULA ITERATION

While a decimal search will tend to the solutions of an equation there are more systematic and efficient approaches that can be taken. An **iterative formula** is simply a formula that will produce a sequence of numbers that converge to a particular value or may diverge. These formulae are defined in the form

$$x_{n+1} = f(x_n)$$

This defines x_{n+1} (the next number in the sequence) in terms of the previous number x_n.

Example 5B

Find the first 5 terms of the sequence defined by

$$x_{n+1} = \frac{1}{2}\left(x_n + \frac{2}{x_n}\right)$$

if $x_1 = 1$.

Solution

We have $x_1 = 1$. The next term x_2 can be obtained by substituting x_1 into the iterative formula to give

$$x_2 = \frac{1}{2}\left(x_1 + \frac{2}{x_1}\right) = \frac{1}{2}\left(1 + \frac{2}{1}\right) = 1.5 .$$

Similarly for x_3, x_4 and x_5

$$x_3 = \frac{1}{2}\left(x_2 + \frac{2}{x_2}\right) = \frac{1}{2}\left(1.5 + \frac{2}{1.5}\right) = 1.417 \quad \text{(to 3 dp)}$$

$$x_4 = \frac{1}{2}\left(1.417 + \frac{2}{1.417}\right) = 1.414 \qquad \text{(to 3 dp)}$$

$$x_5 = \frac{1}{2}\left(1.414 + \frac{2}{1.414}\right) = 1.414 \qquad \text{(to 3 dp)}$$

The iterative formula in Example 5B defines a sequence that converges to a value of 1.414 correct to 3 decimal places.

Consider the general problem of solving $f(x) = 0$. It is always possible to formulate an iterative formula by rearranging the equation to be solved into the form $x = g(x)$. For example consider the equation

$$x^3 + x - 3 = 0$$

One possible rearrangement is

$$x = 3 - x^3$$

and would lead to the iterative formula

$$x_{n+1} = 3 - x_n^3$$

Another possibility is

$$x^3 = 3 - x$$

$$x = \frac{3 - x}{x^2}$$

which would lead to the iterative formula

$$x_{n+1} = \frac{3 - x_n}{x_n^2}$$

Another possibility is

$$x^3 = 3 - x$$

$$x^2 = \frac{3 - x}{x}$$

$$x = \sqrt{\frac{3 - x}{x}}$$

which would give an iterative formula

$$x_{n+1} = \sqrt{\frac{3 - x_n}{x_n}}$$

Which of these iterative formulas leads to the solution?   The TI-92 plot of $x^3 + x - 3$, in Figure 5.7, shows that there is a solution close to $x = 1$.

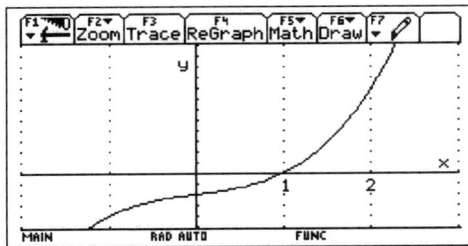

Figure 5.7   Graph of $x^3 + x - 2$ showing root near $x = 1$

Table 5.3 below gives a comparison of each of the 3 formulae all starting with $x_1 = 1$.

Table 5.3

	$x_{n+1} = 3 - x_n^3$	$x_{n+1} = \dfrac{3 - x_n}{x_n^2}$	$x_{n+1} = \sqrt{\dfrac{3 - x_n}{x_n}}$
x_1	1	1	1
x_2	2	2	1.41421
x_3	-5	0.25	1.05892
x_4	128	44	1.35391
x_5	-2097149	-0.0212	1.10263

Here the first two formulae behave in a very strange way, but the final version seems that it may converge. After a number of further iterations (or steps) it converges to 1.21341. So when using formula iteration it is important to be careful as not all iterative formulae will converge.

TI-92 ACTIVITY 5B

This activity will introduce you to iteration on the TI-92. The TI-92 can help you rapidly apply iterative formulae. Clear the HOME screen. Select APPROX mode.

(A)   (i)   Consider the expression $\dfrac{1}{2}\left(x + \dfrac{5}{x}\right)$.

Starting with $x = 2$.

> Type 2 **ENTER**.
> Type (1/2) (ans(1) + 5/ans(1)) **ENTER**

The TI-92 gives 2.25 for the solution.

> Press **ENTER** again to give 2.23611
> Press **ENTER** again to give 2.23607

Continue to show that the sequence converges rapidly.

(ii)   Now try the iterative formula $\sqrt{\dfrac{(3-x_n)}{x_n}} = x_{n+1}$.

Clear the HOME screen.
Starting with 1,

Type 1 **ENTER**
Type $\sqrt{((3 - \text{ans}(1))/\text{ans}(1))}$

Continue the iteration by pressing enter.   Show that this sequence converges slowly.

(B)   This section considers finding the solutions of $x^2 + x - 10 = 0$ using iterative formulae.

(i)   Graph $x^2 + x - 10$, and verify that one solution is close to -4 and the other is close to 3.

(ii)   By hand verify that each of the following iterative formulae can be formulated from the equation $x^2 + x - 10 = 0$.

(a) $x_{n+1} = \sqrt{10 - x_n}$ (c) $x_{n+1} = \dfrac{10 - x_n}{x_n}$

(b) $x_{n+1} = 10 - x_n^2$

(iii)   Type 3 **ENTER**
Type $\sqrt{(10 - \text{ans}(1))}$ **ENTER**

Obtain the first 10 iterations starting with $x = 3$.
Repeat starting with $x = -4$.
What do you notice?

(iv)   Repeat (iii) for each of the other iterative formulae.

(C)   Use the plot facility and the iteration technique together with a suitable rearrangement and iterative formula to find solutions of

(i) $x^2 - 4x + 1 = 0$

(ii) $x^3 + 2x - 5 = 0$

(iii) $x^3 - 3x^2 + 4x - 2 = 0$

1. Use the iterative formula

$$x_{n+1} = 7 - \frac{5}{x_n}$$

starting with $x_1 = 6$ to find a solution of $x^2 - 7x + 5 = 0$ correct to 4 decimal places. Can you use this iterative formula to find the root close to $x = 1$?

2. Use the iterative formula

$$x_{n+1} = \frac{x_n^2 + 5}{7}$$

starting with $x_1 = 0$ to find a solution of $x^2 - 7x + 5 = 0$ correct to 4 decimal places. Can you use this iterative formula to find the root close to $x = 6$?

3. (a)   Show that

$$x_{n+1} = \sqrt{\frac{30}{x_n}} \quad \text{and} \quad x_{n+1} = \frac{2x_n}{3} + \frac{10}{x_n^2}$$

are two iterative formula that could be derived from the equation $x^3 - 30 = 0$.

(b)   Using $x_1 = 3$ find x_5 for both methods.

(c)   Comment on your results and find $\sqrt[3]{30}$ correct to 4 decimal places.

4. The iterative sequence

$$x_{n+1} = \frac{1}{2}\left(x_n + \frac{a}{x_n}\right)$$

is well known for finding solutions of the equation $x^2 = a$.

(a)   Use the formula to solve $x^2 = 8$ and $x^2 = 10$.

(b)   Show that $x = \dfrac{1}{2}\left(x + \dfrac{a}{x}\right)$ can be rearranged to give $x^2 = a$.

(c)   For what equation would the iterative formula

$$x_{n+1} = \frac{1}{3}\left(x_n + \frac{a}{x_n}\right)$$

give the solution.   Use it with $a = 10$ and $x_1 = 2$ to verify your prediction.

5.   (a)   Sketch a graph of $y = x^2 - \sin(x)$ and state the value of the smallest root of $x^2 - \sin(x) = 0$.

(b)   Find two rearrangements of the equation $x^2 - \sin(x) = 0$ and test them to see if they converge.

(c)   Find the other root of the equation correct to 3 decimal places.

6.   Use suitable rearrangements to develop iterative formulae to solve the equations below. Find each solution correct to 2 decimal places.

(a) $x^3 + x - 7 = 0$

(b) $x - \cos x = 0$

(c) $e^x - x - 2 = 0$

6

Differentiation

6.1 INTRODUCTION

In many applications of mathematics we are interested in how a quantity is changing. For example, in waiting for a cup of coffee to cool the **rate of change** of its temperature is important; the state of the economy might be measured by the **rate of change** of the money supply; the increase in the number of species of an animal could be measured by the **rate of change** of the population.

In each of these examples we need to define a function to model the situation and then find how the function is changing. In Chapter 2 you investigated the rates of change of exponential functions $f(x) = a^x$ for different values of a by drawing tangents to the graphs of the functions and finding their slopes. For example we found that when $a = 3$ the tangent slopes were larger than the function values and when $a = 2$ the tangent slopes were less than the function values. In this way you discovered the "special exponential" function e^x whose tangent slope is equal to the function value at each point.

In this chapter we link these two ideas together by defining the rate of change of a function to be the slope of the tangent to the graph of the function and calling it the **derivative** of the function. **Differentiation** is the name given to the process of finding the derivative of a function. The rules of differentiation provide an algebraic method of finding rates of change instead of the graphical approach of Chapter 2.

6.2 GRADIENTS AND RATES OF CHANGE

In Chapters 1 and 2 the rate of change of the exponential graphs at any point were described by the gradient of the tangent line to the graph at that point. For the function e^x we found that the rate of change of the graph at a point, P say, was equal to the function value at P. This result makes e^x a rather special function in mathematics.

In the next TI-92 Activity we explore the rate of change of polynomial functions.

TI-92 ACTIVITY 6A

The aim of this Activity is to explore the rate of change of powers of x.
Clear the Y = Editor and GRAPH screen. Select EXACT mode.

(A)   Set up $y1 = x^2$ and graph x^2 for the window

$$xmin = -1, xmax = 4, xscl = 0.5,$$
$$ymin = -2, ymax = 20, yscl = 1.0$$

The tangent line to a graph at a given point can be drawn using **F5** and
selecting A:Tangent.
Respond to 'Tangent at ?' with 1. Write down the slope of the tangent to $y = x^2$
at $x = 1$.
Draw the tangent to x^2 at $x = 2$.
Write down the slope of the tangent at $x = 2$.
The screen dumps in Figure 6.1 show these tangents and their slopes.

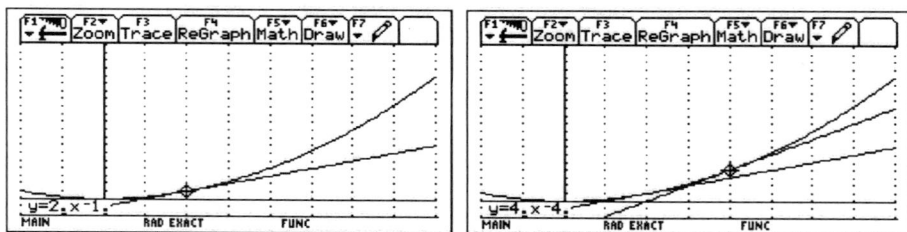

rate of change at $x = 1$ is 2   rate of change at $x = 2$ is 4

Figure 6.1   Tangent lines to $y = x^2$

Now complete the following table.

Table 6.1

x	$y = x^2$	slope of tangent
0		
1	1	2
1.5		
2	4	4
2.5		
3		

Enter the pairs of x and slope values into a Data table (using **APPS** 6:Data/Matrix Editor). Draw a graph of this data.
Find a rule for the slope of the tangent in terms of x.

(B)   Repeat the activity in part (A) for the following functions. In each case find a rule for the slope of the tangent in terms of x.

(i) x^3,   (ii) x^4,   (iii) x^5 (iv) x^6,   (v) x^7

Now complete the following table.

Table 6.2

function	rule for slope of the tangent
x^2	
x^3	
x^4	
x^5	
x^6	
x^7	

What pattern can you see emerging?

(C)   In the same way explore each expression below:

(i) x^{-1},   (ii) x^{-2},   (iii) x^{-3},   (iv) $x^{0.5}$,   (v) $x^{1.5}$

Do you see the same rule for the slope of the tangent that emerged in part (B)?

Definition

The slope of the tangent to the graph of $y = f(x)$ measures the rate of change of $f(x)$ with respect to x. The rule for the slope of the tangent defines the **gradient function** or the **derivative of $f(x)$**. For example, you found that the rule for the slope of the tangent to $f(x) = x^2$ was the function $g(x) = 2x$. We say that $g(x) = 2x$ is the gradient function (or derivative) of x^2.

From what you have done in this TI-92 activity complete the following rule:

gradient function or derivative of $x^n = ?$

(D) Now use the graphical approach to find the gradient function for each of the following expressions:

(i) $3x^2$,　(ii) $5x^6$,　(iii) $12x^{-1}$,　(iv) $14x^{\frac{1}{2}}$

Deduce a rule for finding the gradient function for cx^n where c is a constant.

Summary

Two important features of the gradient or rate of change of x^n have been introduced in this TI-92 Activity. Firstly the slope of the tangent leads to a new function called the gradient function or derivative. Secondly, the rule for the gradient function (or derivative) of powers of x is

$$\frac{d}{dx}(x^n) = nx^{n-1}$$

Furthermore in the last part of the investigation you saw that the derivative of cx^n is cnx^{n-1} where c is a constant.

The notation for the derivative of a function f is written as

$$\frac{d}{dx}(f(x))$$

For example, if $f(x) = x^2$ we write the derivative of x^2 as

$$\frac{d}{dx}(x^2)$$

The rate of change of a function at the value $x = a$	= slope of the tangent to the graph of the function at $x = a$

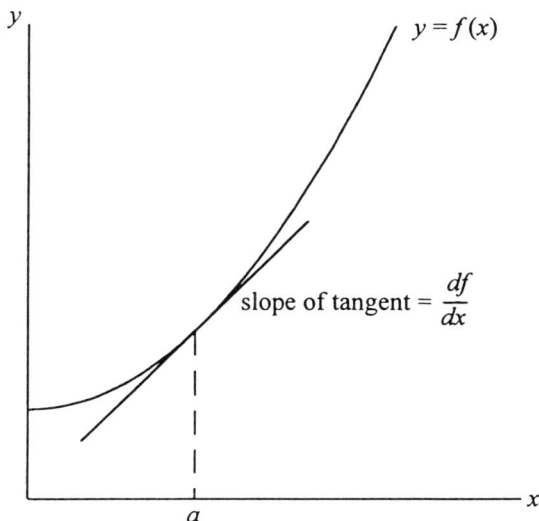

Figure 6.2   Rate of change and the gradient function

Apply these rules in the following exercise.

Exercise 6A

1.   Find the derivative of the following functions.

(a) x^7 (b) $x^{1/3}$ (c) x^{-3} (d) $x^{-5/2}$

(e) $4x^3$ (f) $6x^{3/2}$ (g) $\dfrac{0.7}{x}$ (h) $11x^9$

(i)   5   (j) $6x^5$ (k) $3x^{-1}$

(l) c, where c is a constant.

2.   Find the values of the slope of the tangent of each given function at the given value of x.

(a) x^3 at $x = 2$ (b) x^5 at $x = 1$,   (c) x^2 at $x = 0.5$

(d) x^4 at $x = -1$ (e) $\dfrac{1}{x}$ at $x = 3$ (f) $x^{1/2}$ at $x = 4$.

Check your results by drawing a graph on your TI-92.

TI-92 ACTIVITY 6B

The aim of this investigation is to explore further the rules of differentiation.
Clear the Y = Editor and GRAPH screen. Select EXACT mode.

(A)   Set up $y1 = x^2 + x$ and graph $x^2 + x$ for the window

$$xmin = -1, xmax = 4, xscl = 0.5$$
$$ymin = -2, ymax = 20, yscl = 1$$

Find the slope of the tangent line to the graph at the x-values: $x = 0$, $x = 1$, $x = 2$, $x = 3$ and $x = 4$.
Enter the pairs of x and slope values into a Data table (using **APPS** 6:Data/Matrix Editor). Draw a graph of the data and deduce the equation of the line. Figure 6.3 shows the screen dump for this activity.

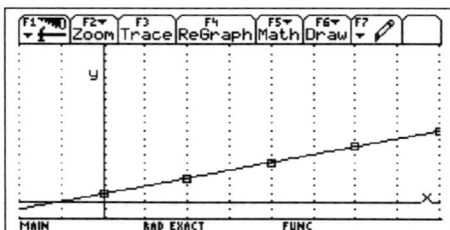

(a)   Table of slope values   (b)   Graph of gradient function

Figure 6.3   Finding the gradient function for $x^2 + x$

The equation of the gradient function is $2x + 1$.

(B)   Repeat the activity in (A) finding the gradient function for each of the following functions. Complete the Table 6.3.

Table 6.3

function	gradient function
$x^2 + x$	$2x + 1$
$x^2 - x$	
$x^3 + x$	
$x^3 + x^2$	
$x^3 - x^2$	

What pattern can you see emerging?

Suppose that you have two functions $u(x)$ and $v(x)$ where $u(x)$ and $v(x)$ are powers of x. Deduce a rule for finding the derivative of $u(x) + v(x)$ and $u(x) - v(x)$.

(C) Use your rule to find the gradient function of $y = 3x^4 - 2x^2 + x$.
Calculate the value of the gradient function at $x = 1$ and $x = 2.5$.
Draw the graph of $y1 = 3x^4 - 2x^2 + x$ and find the slope of the tangent at each point $x = 1$ and $x = 2.5$.
Check that your values of the gradient function are equal to the slope of the tangent.

(D) Repeat the activity in (C) for the following functions and x-values.

  (i) $y = 4x^3 + 3x$ $x = 0, x = 1.5$

  (ii) $y = x^5 - 2x^3 + 1$ $x = -1, x = 1$

  (iii) $y = 0.7x^3 + 0.4x^2 - 0.1x$ $x = 0, x = 2$

(E) The TI-92 provides the gradient function of derivative at the press of a key! This can be done in one of two ways. Select the HOME screen.

  (i)   **F3** select 1:d(differentiate   (see Figure 6.4 (a))

  (ii)   **2nd 8**

Both lead to the result d(in the Entry Line as shown in Figure 6.4 (b).

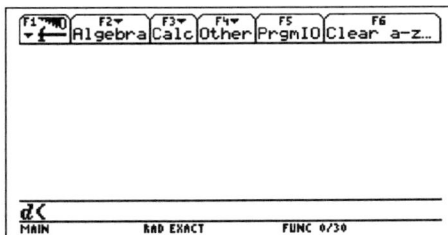

(a) Using **F3**   (b) The Entry Line

Figure 6.4   Differentiating on the TI-92

Now we type in the function ,x) and **ENTER**. Figure 6.5 shows the HOME screen after pressing **ENTER** for the derivative of $3x^4 - 2x^2 + x$.

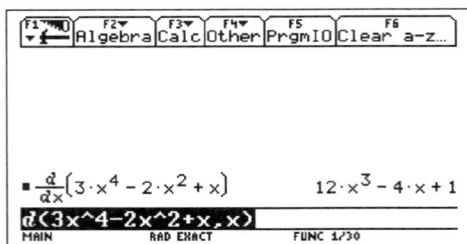

Figure 6.5   Differentiating $3x^4 - 2x^2 + x$

Use your TI-92 to find the derivative of the following functions:

(i) x^5 (ii) x^{-2} (iii) $x^{0.7}$

(iv) $6x^2 - 5x - 1$ (v) $6x^3 - 4x^2 + 3x - 2$ (vi) $8x^9 + 9x^{-2} + 18x^{-3}$

(vii) $\sqrt{x}$ (viii) $0.4x^{1.7} - 0.1x^3$ (ix) $7x^{-2/5} + 4x^{5/2}$

Has the TI-92 given the answers you would have expected from the rule $\dfrac{d}{dx}(x^n) = nx^{n-1}$? If not, can you reconcile the two sets of answers?

(Hint: remember the index laws, $x^{-n} = \dfrac{1}{x^n}$.)

Example 6A

Find the equation of the tangent to the graph of the function $4x^3 + 2x - 8$ at the point $(1,-2)$.

Solution

The tangent is a straight line and the general equation of a straight line is $y = mx + c$. The slope of the tangent m at the point $(1,-2)$ equals the value of the derivative of the function at $x = 1$.

$$\frac{d}{dx}(4x^3 + 2x - 8) = 12x^2 + 2$$

and when $x = 1$, the derivative $= m = 14$.

The tangent has equation $y = 14x + c$.

Since the tangent passes through the point (1,–2) we have

$$-2 = 14 + c \qquad c = -16$$

The equation of the tangent is $y = 14x - 16$.

Exercise 6B

In this exercise use the rules for differentiation (by hand) to find the derivatives. Check your answers with the TI-92 at the end of each problem.

1. Find the derivative of the following functions:

    (a) $6x^5 - 2x + 7$ (b) $3x^{-1} + 2x^{-2}$ (c) $15x - 3x^4 + 2x^7$

    (d) $0.1x^{-3} + 1.9x^3$ (e) $4x^{0.2} - 5x^{-0.1}$ (f) $0.1x^{-3} + 1.9x^{-2}$

    (g) $0.9x^2 - 0.1x^4$ (h) $4.2x^{-1} + 2.1\sqrt{x}$ (i) $\sqrt[3]{x} + \sqrt{x}$

2. Find the slope of the tangent to the graphs of each of the following functions at the given values of x. Hence find the equations of the tangent at each point.

    (a) x^2 at $x = 3$ (b) x^3 at $x = -2$ (c) x^5 at $x = 0.5$

    (d) $x^{1/2}$ at $x = 4$ (e) $x^{5/4}$ at $x = 1$ (f) $x^{0.7}$ at $x = 4$

    (g) x^{-1} at $x = 0.3$ (h) $x^{-1.9}$ at $x = 1.2$ (i) $3x^4$ at $x = -2$

    (j) $2x^{2.5}$ at $x = 9$ (k) $3x^2 - 2x + 1$ at $x = 0$ (l) $5x^{-1} - 2x$ at $x = 1$

3. Find the slope of the tangent to the graphs of each of the following functions at the given values of x. Deduce whether the functions are increasing or decreasing at the given values.

    (a) $4x^3 + 2x^2 - x$ $x = 0, x = 1.$

    (b) $x^3 - 3x^2 + 3x - 1$ $x = -1, x = 1, x = 2.$

    (c) $5x^{-1} - 2x$ $x = 1, x = 2.$

    (d) $x^{1.3} + x^{-0.7}$ $x = 0.5, x = 1.5.$

    What can you deduce about the shape of the graph between $x = 0.5$ and $x = 1.5$.

6.3   PHYSICAL SITUATIONS AND RATES OF CHANGE

In the last two sections we have introduced the idea of the rate of change of a graph and defined the gradient function or derivative. From the graphs we have deduced basic rules for differentiating powers of x and sums/differences of powers of x.

Now we turn to an algebraic view of rates of change and in this section we introduce a formal definition of differentiation. The idea of a rate of change as a speed or an acceleration is probably familiar to you. As an example of the ideas involved consider a recent journey between two of the campuses of the University of Plymouth in Plymouth and Exmouth, a distance of 50 miles. The journey out of Plymouth was pretty slow with a high density of traffic and a speed limit. On the A38 between Plymouth and Exeter we travelled at a steady 70mph only slowing down for road works near Newton Abbot. The last part of the journey from the M5 to Exmouth is along country lanes where once again speed is restricted. The journey took 1¼ hours. What can we say about the speed of the journey? Clearly the speed given by the speedometer changed many times (probably continuously) during the trip. It would have been 0mph at traffic lights and up to 70mph on the motorway. Looking at speed more globally, since the 50 mile journey took 1¼ hours, the *average speed* is just 50/1¼ = 40mph. An average speed could be used to give us a rough idea of the time for a journey. Suppose I travel 100 miles along similar roads to the trip above. I might expect an average speed of 40mph and a journey time of 2½ hours. The speedometer would give some idea of the speed at an instant so this is called *the instantaneous speed*. (It is this speed that the police would be interested in if you travel at 70mph through a built up area!!). Since

$$\text{average speed} = \frac{\text{distance travelled}}{\text{time taken}}$$

we define this as the *average rate of change* of distance with time. Over a very short interval of time the average speed will provide a good approximation to the instantaneous speed. The smaller the time interval the better the approximation. In mathematical language we would write

$$\text{instantaneous speed} = \lim_{t \to 0} \frac{s}{t}$$

where s is the distance travelled in time t. A natural question is what does this mean? As t gets smaller so does s and in the limit we have 0/0! This leads us into the ideas of limits again.

MATHEMATICAL INVESTIGATION 6A

A small object slides along a table so that the distance from its starring point (in metres) is modelled by

$$s = -1.3t^2 + 7.8t$$

where t is the time in seconds.

The object comes to rest after 3 seconds.

(i)   Calculate the total distance travelled and the average speed for the journey.

(ii)   Calculate the average speed during the 1st second ($t = 0$ to 1), during the 3rd second of the journey ($t = 2$ to 3).

(iii)   By completing the following table find the instantaneous speed when $t = 1$.

Table 6.4

t	s	distance travelled between $t = 1$ and t	average speed between $t = 1$ and t
1.2	11.232	$11.232 - 6.5 = 4.732$	$4.732/0.2 = 23.66$
1.1			
1.01			
1.001			
1.0001			

Estimate the instantaneous speed at $t = 1$.

(iv)   By choosing appropriate times repeat part (iii) to estimate the instantaneous speed at time $t = 2$ seconds.

(v)   Find the value of $\dfrac{ds}{dt}$ at $t = 1$ and $t = 2$.

What can you deduce from your results?

This activity suggests $\displaystyle\lim_{t\to 0}\frac{s}{t}$ does converge and that the instantaneous speed is associated with the derivative of the distance function.

Consider a more formal algebraic approach. Let $s(t) = -1.3t^2 + 7.8t$ and suppose we wish to evaluate the speed at $t = t_0$. For a small time h the distance travelled between $t = t_0$ and $t = t_0 + h$ is

$$s(t_0 + h) - s(t_0) = \left[7.8(t_0 + h) - 1.3(t_0 + h)^2\right] - [7.8t_0 - 1.3t_0^2] = 7.8h - 2.6t_0h - 1.3h^2$$

The average speed during the time interval h is

$$\text{average speed} = \frac{\text{distance travelled}}{\text{time taken}}$$

$$= \frac{7.8h - 2.6t_0h - 1.3h^2}{h} = 7.8 - 2.6t_0 - 1.3h$$

As the value of h decreases, the distance travelled decreases but the average speed tends to $7.8 - 2.6t_0$.

We deduce that the instantaneous speed at time t_0 is $7.8 - 2.6t_0$.

Notice that the derivative of $s(t)$ is $7.8 - 2.6t$ so that at time $t = t_0$

$$\text{instantaneous speed} = \text{rate of change of distance}$$
$$= \text{derivative of distance function}$$

For an object moving so that its distance function is modelled by $s(t)$ at time t, the instantaneous speed of the object v is given by

$$v = \frac{ds}{dt}$$

The acceleration is related to the speed in a similar way. We have

$$\text{average acceleration} = \frac{\text{change in speed}}{\text{time}}$$

and

$$\text{instantaneous acceleration} = \text{rate of change of speed}$$
$$= \lim_{\text{time}\to 0} \frac{\text{change in speed}}{\text{time}} = \frac{dv}{dt}$$

░░░░░░░░░░░░░░ **Exercise 6C** ░░░░░░░░░░░░░░░

1. A car travels from London to Leeds, a distance of 180 miles in 4½ hours. Calculate the average speed for the journey.

    If most of the journey takes place on the M1, suggest reasons why the average speed is not nearer 70 mph.

    Sketch a possible graph for the distance travelled against time assuming that the driver stops at two motorway service stations.

2. The speed of an object rolling on a sloping table changes from 3.1 ms^{-1} to 4.6 ms^{-1} in 1.3 seconds. Calculate the average acceleration of the object during this time interval.

3. A stone is thrown up in the air so that its height above the thrower is modelled by

    $$h = 40t - 5t^2$$

    where t is time in seconds.

    (a) Find the average speed of the stone during the time intervals

        (a) $t = 0$ to 1   (b) $t = 1$ to 2   (c) $t = 2$ to 2.5.

    (b) Find the instantaneous speed of the stone at times $t = 0$, 1, 2, 3 and 4 seconds. What happens to the stone at $t = 4$?

    (c) Find the instantaneous acceleration of the stone at times $t = 0$, 1, 2, 3 and 4 seconds. What does the negative sign mean?

4. The population of a species of fish in a lake is modelled by

    $$P = 4\sqrt{t} + 3$$

    where t is the time in seconds.

    Calculate the (instantaneous) rate of change of the population at time $t = 2$ and $t = 9$.

5.   The area of an ink blot is given in terms of its radius by

$$A = \pi r^2$$

Find the rate of change of the area with respect to r.

6.   A function f is defined by $f(x) = 3x^2 - 4x + 6$.

   (a)   By completing the following table estimate the instantaneous rate of change of f when $x = 1.5$.

x	$f(x)$	change in $f(x)$ between x and 1.5	average rate of change between x and 1.5
1.7			
1.6			
1.55			
1.51			
1.501			
1.5001			

   (b)   Check your estimate using differentiation.

7.   A function p is defined by $p(r) = 3r - 4/r$.
      By completing the following table estimate the instantaneous rate of change of p when $r = 2$.

r	$p(r)$	change in $p(r)$ between r and 2	average rate of change between r and 2
2.2			
2.1			
2.01			
2.001			
2.0001			

Limit Definition of Differentiation

Consider a general function $f(x)$ and its change between two values of x, $x_0 + h$ and x_0. The average rate of change of f is defined by

$$\text{average rate of change} = \frac{f(x_0 + h) - f(x_0)}{h}$$

On a graph of f this average rate of change is the slope of a chord between points $(x_0, f(x_0))$ and $(x_0 + h, f(x_0 + h))$ (see Figure 6.6).

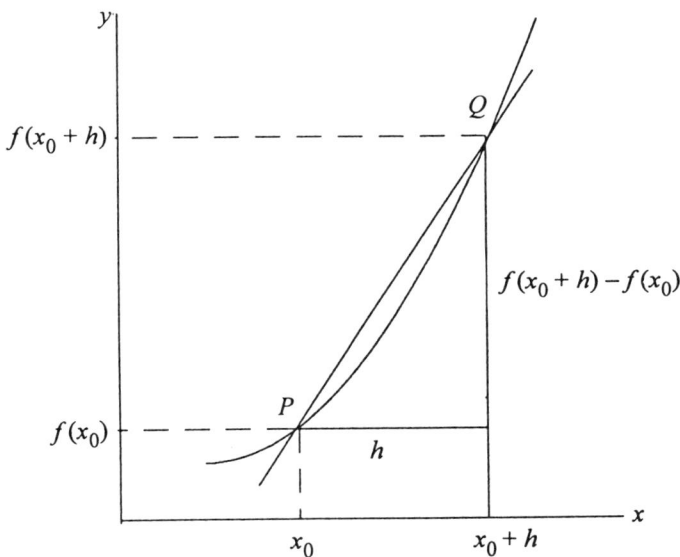

Figure 6.6

As the size of h is reduced the point Q moves round the curve towards point P. The slope of the chord approaches the slope of the tangent to the curve at P (see Figure 6.7).

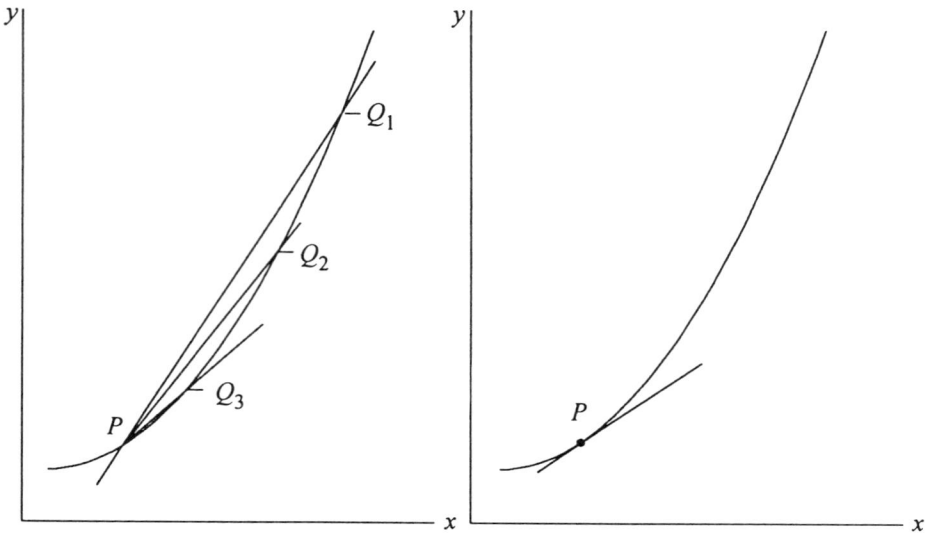

Figure 6.7

We have seen that the slope of the tangent at P is the value of the derivative of the function at P. So now we have a formal link between rates of change and the derivative.

Definition of the derivative of a function $f(x)$

$$\frac{d}{dx}(f(x)) = \lim_{h \to 0} \frac{f(x+h) - f(x)}{h}$$

The notation for the derivative of $f(x)$ with respect to x is often written as $f'(x)$. An alternative geometrical view of the derivative is known as the **local straightness** of the graph of $f(x)$. If you zoom in on a small portion of the graph centred at point P, then the graph looks like a straight line. This line has the same slope as the tangent at P.

TI-92 ACTIVITY 6C

The aim of this investigation is to demonstrate the link between the limit definition and differentiation. Clear the HOME screen.

(A)   In this activity we find the derivative of x^2 using the limit definition

$$\frac{df}{dx} = \lim_{h \to 0} \frac{f(x+h) - f(x)}{h}$$

Use **F3**, select 3:limit (and enter

limit(((x + h) ^ 2 – x ^ 2)/h,h,0)

being very careful with the brackets.

On the screen you should see

$$\lim_{h \to 0} \left(\frac{(x+h)^2 - x^2}{h} \right)$$ and the result $2x$.

What is the derivative of x^2 from the 'gradient function' approach?
We see that the limit definition of algebra gives the same result as the gradient function definition of the graphical approach.

(B)   Now repeat the activity in (A) evaluating the limit and using the rules for differentiating powers to confirm that the algebraic and graphical definitions give the same results.

(i) $\lim_{h \to 0} \dfrac{(x+h)^3 - x^3}{h}$

        $\dfrac{d(x^3)}{dx}$

(ii) $\lim_{h \to 0} \dfrac{(x+h)^4 - x^4}{h}$

        $\dfrac{d(x^4)}{dx}$

(iii) $\lim\limits_{h \to 0} \dfrac{\left\{4(x+h)^3 - 2(x+h)^2 + 7\right\} - \left\{4x^3 - 2x^2 + 7\right\}}{h}$

$\dfrac{d}{dx}(4x^3 - 2x^2 + 7)$

(iv) $\lim\limits_{h \to 0} \dfrac{\left\{13(x+h)^7 - 5(x+h)^4 + 13(x+h)\right\} - \left\{13x^7 - 5x^4 + 13x\right\}}{h}$

$\dfrac{d}{dx}(13x^7 - 5x^4 + 13x)$

(v) $\lim\limits_{h \to 0} \dfrac{\left\{4(x+h)^{-\frac{1}{2}} + \dfrac{3}{(x+h)}\right\} - \left\{4x^{-\frac{1}{2}} + \dfrac{3}{x}\right\}}{h}$

$\dfrac{d}{dx}\left(4x^{-\frac{1}{2}} + \dfrac{3}{x}\right)$

Choose other functions and check that the limit process gives the same result as differentiation.

Example 6B

A function $f(x)$ is given by

$$f(x) = 4x^3 + 2x^2 - 3x - 7$$

By evaluating $\lim\limits_{h \to 0} \dfrac{f(x+h) - f(x)}{h}$ confirm the limit definition of the derivative.

Solution

In this example we carry out the algebra 'by hand'.

$$f(x+h) = 4(x+h)^3 + 2(x+h)^2 - 3(x+h) - 7$$
$$= 4x^3 + 12x^2h + 12xh^2 + 4h^3 + 2x^2 + 4xh + 2h^2 - 3x - 3h - 7$$

$$f(x+h) - f(x) = 12x^2h + 12xh^2 + 4h^3 + 4xh + 2h^2 - 3h$$

Dividing both sides by h,

$$\frac{f(x+h)-f(h)}{h} = 12x^2 + 12xh + 4h^2 + 4x + 2h - 3$$

As h tends to zero this ratio becomes

$$\lim_{h \to 0} \frac{f(x+h)-f(x)}{h} = 12x^2 + 4x - 3$$

Using the rules of differentiation of powers of x we have

$$f(x) = 4x^3 + 2x^2 - 3x - 7$$

$$\frac{df}{dx} = 12x^2 + 4x - 3.$$

Confirming the formal definition of differentiation.

Summary

We now bring together the ideas of differentiation introduced so far.

The **derivative** of a function $f(x)$ with respect to x is written as $\frac{df}{dx}$ and has the following meanings:

1.   the derivative is the (instantaneous) **rate of change** with respect to x;

2.   the derivative is the **slope of the tangent** to the graph of $f(x)$;

3.   formally $\frac{df}{dx} = \lim_{h \to 0} \frac{f(x+h)-f(x)}{h}$;

4.   if $f(x) = x^n$ then $\frac{df}{dx} = nx^{n-1}$.

$\frac{df}{dx}$ is also called the **gradient function** or derived function.

   An alternative notation for the derivative of a function $f(x)$ with respect to x is $f'(x)$. We shall use both notations $\frac{df}{dx}$ and $f'(x)$ in subsequent chapters.

Exercise 6D

In this exercise evaluate each limit 'by hand'.

1. Confirm the link between $\lim\limits_{h \to 0} \dfrac{f(x+h) - f(x)}{h}$ and the rule for differentiating powers of x for the following functions.

   (a) x^2 (b) x

   (c) $3x^2 - 4x$ (d) $5x^2 + 2x - 8$

   (e) x^3 (f) $2x^3 + 3x - 1$

   (g) $\dfrac{1}{x}$ (h) c, where c is a constant.

2. In problem 1 the independent variable is x. What would be the formal limit definition for the derivative of a function $g(t)$ where t is the independent variable? Use your formula for the following functions.

   (a) t (b) $3t^2 - 4t + 1$

   (c) $5t^3 - 6t$ (d) 4

   Repeat each limit using your TI-92 and variable t.

3. For each of the following functions write down its derivative as a formal limit. Then evaluate the limit to find the derivative.

   (a) $f(u) = u^2 - 3u + 5$

   (b) $g(s) = s^3 + s$

   (c) $v(t) = -0.5t^2 + 10t - 1$

   (d) $h(u) = \dfrac{1}{u} + 4u$

   (e) $f(x) = x^5$

6.4   MAXIMUM AND MINIMUM VALUES

Features of Graphs of Functions

Figure 6.8 shows the graphs of two quadratic functions $x^2 - 4$ and $4 - x^2$.

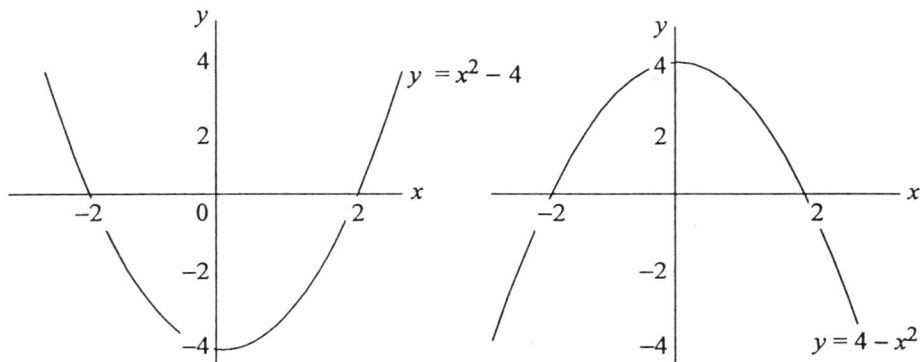

Figure 6.8

If you were asked to describe the graphs describing their essential features you would probably include the following points.

1.   They each cut the x axis at two points $x = -2$ and $x = 2$.

2.   The graph of $f(x) = x^2 - 4$ has a minimum at the point $(0, ^-4)$. It continuously decreases up to the point $(0, -4)$ and then increases as x continues to increase.

3.   The graph of $f(x) = 4 - x^2$ has a maximum at the point $(0,4)$. It increases until the point $(0,4)$ and then decreases as x continues to increase.

To describe the graph of a function we note the features as the direction of x increases, i.e. we scan the graph moving from left to right.

TI-92 ACTIVITY 6D

The aim of this investigation is to explore some important features of a graph of a function using the derivative.
Clear the Y = Editor and the GRAPH screen. Choose AUTO mode.

(A) (i)   Set up $y1 = x^5 - 3.75x^3 - 1.25x^2 + 3.75x + 3.5$ and graph the function using the window

$$xmin = -2, xmax = 2, xscl = 0.2$$
$$ymin = -2, ymax = 5, yscl = 0.5$$

Describe the important features of the graph in words.

Use **F5** and A:Tangent to draw the tangent to the graph at $x = -1.5$, $x = -1.0$, $x = -0.5$, $x = 0.5$, $x = 1.0$, $x = 1.5$ and $x = 2.0$.

Make a note of the slope of the tangent at each of these points and complete the following table.

Table 6.5

x	value of slope	graph is increasing, decreasing or neither
−1.5	7.5	increasing
−1.0		
−0.5		
0.5		
1.0		
1.5		
2.0		

Figure 6.9 shows a TI-92 screen dump for this activity.

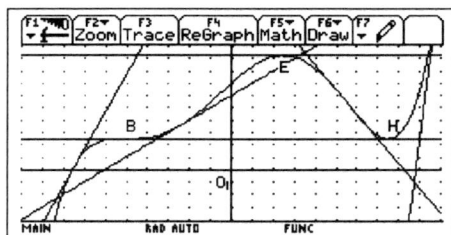

Figure 6.9   Graph and tangents to $y1$

At points, B, E and H the slopes of the tangents are zero. We call these points **turning points** or **stationary points**. At these points $\dfrac{df}{dx} = 0$.

(ii)   Use the **Differentiate** command to find $\dfrac{df}{dx}$ and show that $\dfrac{df}{dx} = 0$ at $x = -1, 0.5$ and 1.5.

The shape of the curve is quite different at points B, E and H.

As you move along the curve from A to B to C, the slope of the tangent is positive at A, zero at B and positive at C. The curve crosses the tangent. Point B is an example of **a point of inflexion**.

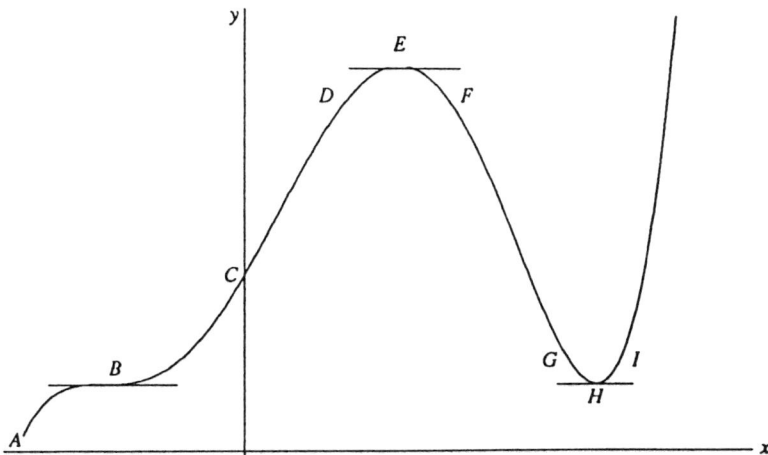

Figure 6.10

(iii)   Discuss whether there is another point of inflexion with the property that the curve crosses the tangent.

At point E the slope changes from a positive value at D to a negative value at F with the rate of change being 0 at E.
Point E is an example of **a local maximum**.

Point H is an example of a **local minimum** at which the slope changes from a negative value at G to a positive value at I with the rate of change being 0 at H.

The TI-92 will locate local maximum, minimum and points of inflexion.

In Graph Screen Press **F5** select 4:Maximum, **ENTER**.

The TI-92 asks for a lower bound, move the cursor to the left of the point E, Press **ENTER**. The TI-92 now asks for an upper bound, move the cursor to the right of point E, press **ENTER**. This gives the x and y value for the local maximum located at $x = 0.5$.

Local minimums and inflections are found similarly by sandwiching the point between a lower and upper bound.

(B)   Clear the HOME and GRAPH screens.   Set up and graph the following functions.

(i) $f(x) = x^3 - 4x$

(ii) $f(x) = 3x^4 + 4x^3 - 24x^2 - 48x - 5$

From your graphs describe the features of each function.   Find the turning points using **F5**, 4:Maximum, 3:Minimum, etc.   Show that at each turning point $\dfrac{df}{dx} = 0$.

(C)   Clear the HOME screen, Y = Editor and GRAPH screen.
Set up and graph the following functions:

$$f(x) = 6 + 4x - x^2 \quad \text{as } y1(x)$$

and $f'(x)$ as $y2(x)$

using the WINDOW

xmin = –4, xmax = 6, xscl = 1
ymin = –6, ymax = 18, yscl = 1 .

Answer these three questions and summarise your answers in the table.

(a)   What is happening to $f'(x)$ when $f(x)$ has a local maximum?

(b)   What is happening to $f(x)$ when $f'(x)$ is positive?

(c)   What is happening to $f(x)$ when $f'(x)$ is negative?

Table 6.6

$f'(x)$	$f(x)$
	local maximum
positive	
negative	

Repeat this activity for the following functions:

(i) $f(x) = x^2 + 4$ xmin = –2, xmax = 2, xscl = 0.5
                                                 ymin = –2, ymax = 10, yscl = 1

(ii) $f(x) = x(x – 1)(x + 1)$ xmin = –2, xmax = 2, xscl = 0.5
                                                 ymin = –2, ymax = 2, yscl = 0.5

(iii) $f(x) = \dfrac{1}{(1+x^2)}$ xmin = –2, xmax = 2, xscl = 0.5
                                                 ymin = –2, ymax = 2, yscl = 0.5

(D)   Clear the Y = Editor and GRAPH screen.
      Graph the following functions and from the graph predict the shape of $f'(x)$ 'by hand' before you graph it with the TI-92.

(i) $f(x) = (x – 1)(x + 2)$ xmin = –2, xmax = 2, xscl = 0.5
                                                 ymin = –4, ymax = 4, yscl = 1

(ii) $f(x) = x(x – 2)(x + 1)(x + 3)$ xmin = –4, xmax = 4, xscl = 1
                                                 ymin = –10, ymax = 10, yscl1 = 1

(E)   Clear the Y = Editor and the GRAPH screen.
       The function $g(x) = x + 1$ is the derivative of some unknown function $f(x)$.
       Graph the function $g(x)$.
       From your graph describe the properties of $f(x)$ and propose a graph of $f(x)$.

       Repeat this activity with the following functions:

       (i) $g(x) = 2 - x$

       (ii) $g(x) = x^2 - 3x + 2$

       Note that in Chapter 7 we will find an algebraic method of finding $f(x)$ when
       $g(x) = f'(x)$ is known. This process is called integration. For the functions
       $g(x)$ above possible functions $f(x)$ are given so that you can compare your
       answers to the form of $f(x)$.

       For $g(x) = x + 1$ $f(x) = \dfrac{x^2}{2} + x - 2$

       For $g(x) = 2 - x$ $f(x) = 1 + 2x - \dfrac{x^2}{2}$

       For $g(x) = x^2 - 3x + 2$ $f(x) = \dfrac{x^3}{3} - \dfrac{3x^2}{2} + 2x$

Example 6C

Find the turning points of the function $f(x) = x^4 - 4x^2 + 2$ and describe them.

Solution

The TI-92 would provide a graphical means of describing the properties of this
function. In this example we will use an algebraic approach and calculus.

The turning points are given by $\dfrac{df}{dx} = 0$.

Differentiating f we have

$$\frac{df}{dx} = 4x^3 - 8x = 4x(x^2 - 2)$$

Solving for $4x(x^2-2) = 0$ for x gives:

$$x = 0 \text{ or } x = -\sqrt{2} \text{ or } x = \sqrt{2}$$

There are three turning points. To classify them we look at the slope of the tangent (i.e. the value of the derivative) either side of the turning point.

x	$\dfrac{df}{dx}$	shape of graph	classification
-1.5 $-\sqrt{2}$ -1	-1.5 0 4		local minimum
-0.5 0 0.5	3.5 0 -3.5		local maximum
1 $\sqrt{2}$ 1.5	-4 0 1.5		local minimum

Figure 6.11 shows the graph of the function confirming the results in the table.

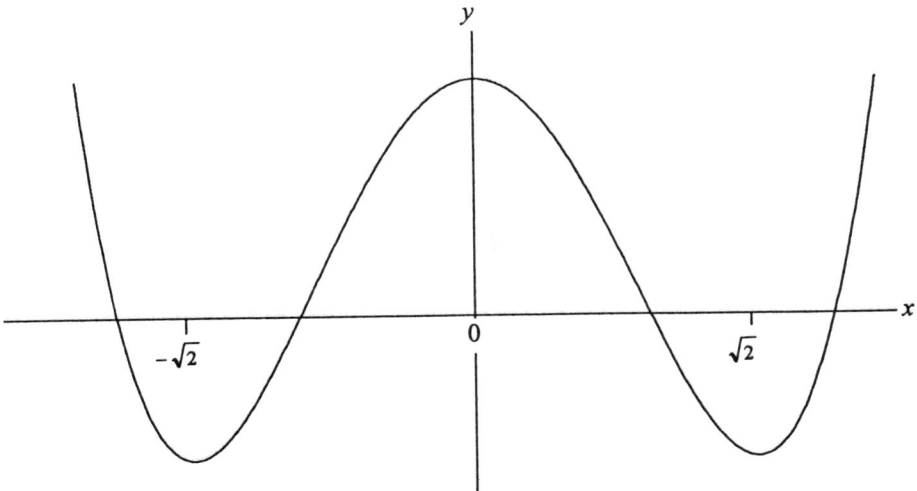

Figure 6.11

░░░░░░░░░░░░ **Exercise 6E** ░░░░░░░░░░░░░░░░░░░░░░░░░░░░░░░░░░░░

1.   Find the turning points of the following functions and classify them. Sketch
      the graphs of the functions to confirm your results.

      (a) $3x^2 - 12$ (b) $9 - x^2$

      (c) $2x^2 + 5x - 3$ (d) $2 - x - 3x^2$

      (e) $x^3 + 3$ (f) $x^3 - 5x^2 - x + 5$

      (g) $5 - 12x - 2x^2 + 4x^3 - x^4$ (h) $x + \dfrac{1}{x} - 2$

2.   Use the TI-92 to draw a graph for each of the functions in problem 1.
      Use your graphs to investigate whether any of the functions have a point of
      inflexion.

3.   Consider the following graphs of two functions.
      Show where the function is

      (a) increasing,   (b) decreasing,   (c) stationary.

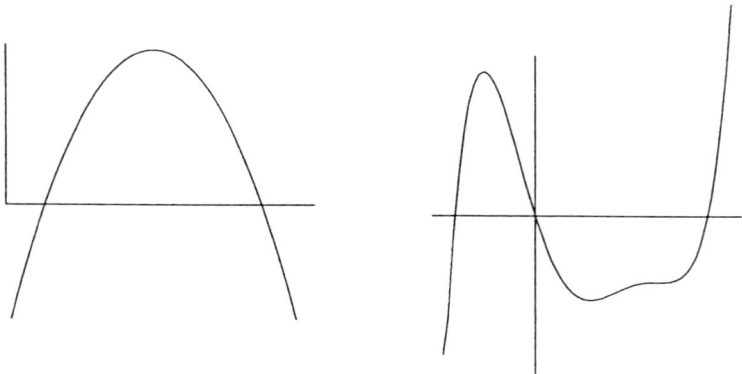

Figure 6.12

4.   You are given the following information about a function $f(x)$

      (i) $f(x) = 1$ when $x = 0$;

      (ii)   at the points (1,2), (3,–2) and (5,3) the graph is "stationary";

(iii) $f(x) = 0$ when $x = -\frac{1}{2}$, when $x = 2$, when $x = 4$ and when $x = 6$;

(iv)   the gradient is positive when $x = 0$ and when $x = 4$;

(v)   the gradient is negative when $x = 2$ and when $x = 6$.

Use all this information to draw a rough sketch graph of the function.

The Second Derivative

When a function f is differentiated a new function is obtained. This new function is called the derived function or gradient function. For example, if $f(x) = x^3 - 3x^2 + 1$ the derived function is

$$\frac{df}{dx} = 3x^2 - 6x$$

We can differentiate this function to obtain

$$\frac{d}{dx}(3x^2 - 6x) = 6x - 6$$

So we can think of the function $6x - 6$ as the result of having differentiated the initial expression $x^3 - 3x^2 + 1$ twice. This new expression $6x - 6$ is called the **second derivative** of the function f and is written as

$$\frac{d}{dx}\left(\frac{df}{dx}\right) = \frac{d^2f}{dx^2} = f''(x)$$

To find the second derivative using the TI-92 we type $d(f(x),x,2)$. Figure 6.13 shows the second derivative of $x^3 - 3x^2 + 1$.

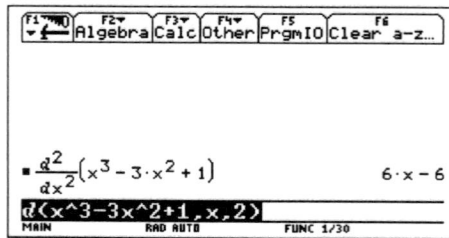

Figure 6.13   Finding second derivatives on TI-92

Example 6D

Find the first derivative and second derivative of the function $f(x) = x^4 - 3x^3 + 4x - 7$.

Solution

The first derivative is just the derived function

$$\frac{df}{dx} = 4x^3 - 9x^2 + 4$$

To find the second derivative we just differentiate again

$$\frac{d^2f}{dx^2} = \frac{d}{dx}(4x^3 - 9x^2 + 4) = 12x^2 - 18x$$

Exercise 6F

1.   Find the second derivative of each of the following functions 'by hand'.

(a) $x^3 - 3x$

(b) $4x - 2x^3 + 5x^5 - x^6$

(c) $x^2 - x^{\frac{1}{2}} + x^{-\frac{1}{4}}$

(d) $x + \dfrac{1}{x}$

(e) $2\sqrt{x} + x^3$

(f) $3x + 4x^2 - 7x^6$

(g) $4t^2 - 3t + 7$

(h) $11t - 4$

(i) $t^{\frac{1}{2}} - 2t^{\frac{1}{4}} + 3t^{-\frac{1}{3}}$

(j) x^n where n is a constant.

2.   Find the second derivatives of the functions in problem 1 using the TI-92.

3. We can continue the process, so that the third derivative of f is the derivative of $f''(x)$ and is written $f'''(x)$. Find the second, third and fourth derivatives of the following functions.

(a) $x^5 - 2x^4 + 3x$ (b) $4x - 2x^3 + 5x^5 - x^6$

(c) $\dfrac{1}{x}$ (d) $\sqrt{t} - 2t^3$

Check your answers using the TI-92.

TI-92 ACTIVITY 6E

The aim of this investigation is to relate values of the second derivative of a function f to the graph of the function f. Clear the Y = Editor, HOME and GRAPH screens.

(A) Split the screen using **MODE F2** and select LEFT-RIGHT.
For 'Split 1 App' choose Y = Editor and for 'Split 2 App' choose Graph.
Define the window:

> $x$min $= -2$, $x$max $= 3$, $x$scl $= 0.2$
> $y$min $= -4$, $y$max $= 5$, $y$scl $= 0.5$

(i) Set up $y1 = x^3 - 3x^2 - x + 3$ and $y2 = d(y1(x),x)$.
Graph $y1$ and $y2$.
Your TI-92 screen should look like Figure 6.14.

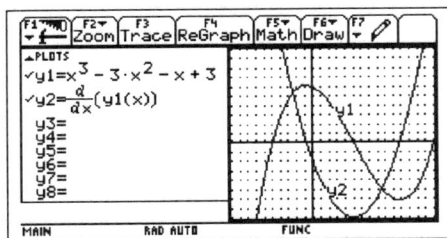

Figure 6.14 Graphs of $y1(x)$ and $\dfrac{dy1(x)}{dx}$ Figure 6.15 Graphs of $y1(x)$ and $\dfrac{d^2 y1(x)}{dx^2}$

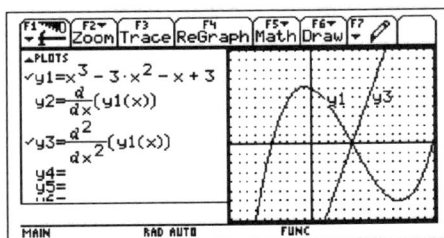

(ii) Set up $y3 = d(y1(x),x,2)$, i.e. the second derivative of $y1(x)$.
Graph $y1$ and $y3$. ('Turn off' the function $y2$.)
Your TI-92 screen should look like Figure 6.15.

Check the following statements from your graph.

1.   The zeros of the derivative coincide with the local maximum and local minimum of f.

2.   At the local maximum the second derivative is negative.

3.   At the local minimum the second derivative is positive.

4.   At the point of inflexion the second derivative is zero.

(B)   Investigate these statements with the following functions.

(i) $f(x) = x^3 + x^2 - 6x$ (iii) $f(x) = x^4$

(ii) $f(x) = x^4 - 5x^2 + 4$ (iv) $f(x) = -2x^3 + x^2 + 7x - 6$

Discuss how values of the second derivative can be used to explore the properties of a function.

(C)   Figure 6.16 shows graphs of $f(x) = x(x-1)(x+2)$, $f'(x)$ and $f''(x)$.

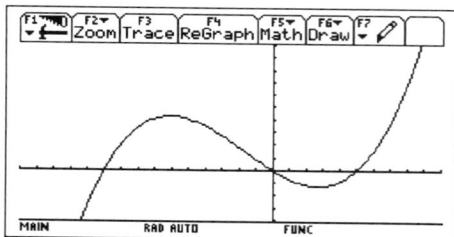

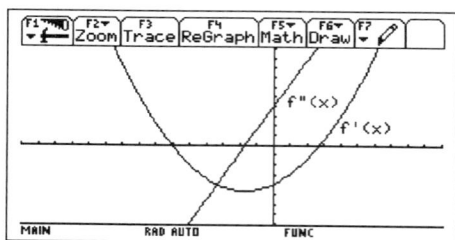

Graph of $f(x)$ Graphs of $f'(x)$ and $f''(x)$

Figure 6.16

Discuss how you might partially reconstruct $f(x)$ if you only knew the graph of $f''(x)$.

Try out your ideas using the functions.

(i) $f''(x) = 6x - 2$ for which $f(x) = x^3 - x^2$

(ii) $f''(x) = 12x^2$ for which $f(x) = x^4 + x - 1$

(D)   At a point of inflexion the second derivative is zero.
Find the second derivatives at $x = 0$ of the following functions.

(i) $f(x) = x^3$ (ii) $f(x) = x^4$

(iii) $f(x) = x^5 - 3$ (iv) $f(x) = 2 - x^6$

Graph each function. Describe the behaviour of the graph when $x = 0$.
Does $f''(x) = 0$ always give a point of inflexion?
Clearly you need to be very careful when dealing with points of inflexion.

Summary

The second derivative represents the rate of change of the tangent to the graph of a
function, and its value at the turning points can be used to classify them.
    At a local maximum:

$$\frac{df}{dx} = 0 \quad \text{and} \quad \frac{d^2 f}{dx^2} < 0$$

(the slope of the tangent is decreasing from positive to negative values and so the rate
of change of tangent is negative).
    At a local minimum

$$\frac{df}{dx} = 0 \quad \text{and} \quad \frac{d^2 f}{dx^2} > 0$$

(the slope of the tangent is increasing from negative to positive values and so the rate
of change of the tangent is positive).
    At a point of inflexion $\frac{d^2 f}{dx^2} = 0$. If the first derivative is also zero at a point of
inflexion the graph may look like Figure 6.17.

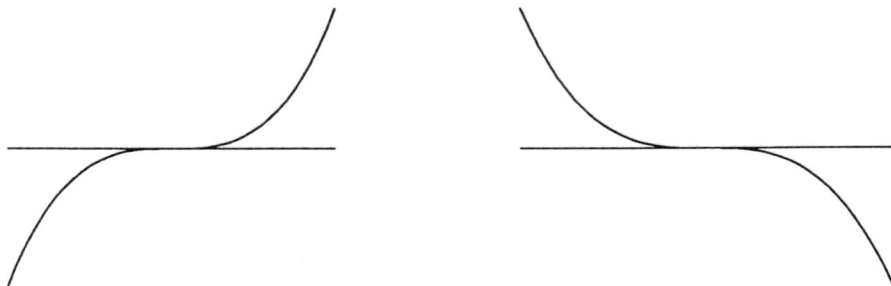

Figure 6.17

You should use these results with care when classifying the turning points of a function. For example, consider the functions $f(x) = x^3$ and $f(x) = x^4$.

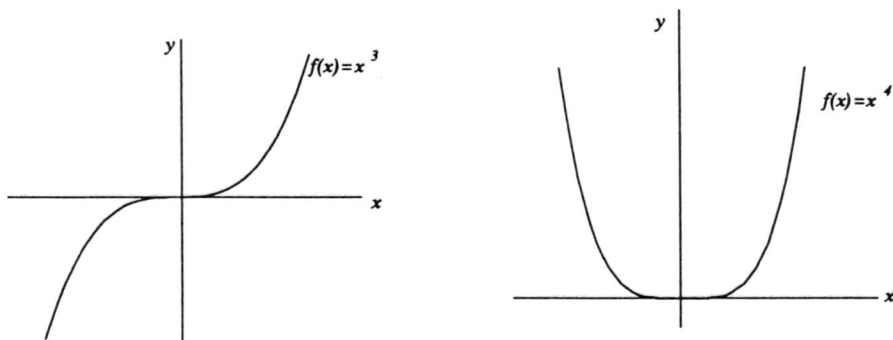

Figure 6.18

For $f(x) = x^3$, $\dfrac{df}{dx} = 0$ and $\dfrac{d^2 f}{dx^2} = 0$ at $x = 0$.

For $f(x) = x^4$, $\dfrac{df}{dx} = 0$ and $\dfrac{d^2 f}{dx^2} = 0$ at $x = 0$.

However notice the different shapes of the two graphs; $f(x) = x^3$ has a point of inflexion at $x = 0$ whereas $f(x) = x^4$ has a local minimum at $x = 0$.

If the first and second derivatives of a function are both zero at a value of x, it is necessary to look at the slope of the tangent either side of the turning point in order to classify it.

Find the coordinates of the turning points of the following functions.   Where appropriate use the second derivative to classify them.

(a) $4x^2 - 11$

(b) $5x^3 - 15x^2 + 15x + 2$

(c) $x^4 - 6x^3 + 11x^2 - 6x$

(d) $x^3 - 6x^2 + 11x - 6$

(e) $x + \dfrac{1}{x}$

(f) $x^2 - \dfrac{1}{x^2}$

Use the TI-92 to sketch the graphs of the functions to check your answers.

Applications

In this section we have explored the application of the derivative to investigate the properties of the graphs of functions. The ideas are also useful in solving problems involving the maximum and minimum values of physical quantities.

Example 6E

A carton is being designed to hold ½ litre of milk. The carton must have a square base to fit neatly onto supermarket shelves.

Find the dimensions of the carton so that the material to be used is minimised.

Solution

Step 1   Formulate the mathematical problem

The first step in problems of this type is to set up the function to be minimised (or maximised in other problems). Assume that the carton is to be a cuboid with base size x and height h (in cm).

Volume of box $= x^2 h$ cm^3

The volume of ½ litre of milk is 500 cm^3.

Hence

$$x^2 h = 500 \qquad (1)$$

The area of material to be used is

$$A = 2x^2 + 4xh \qquad (2)$$

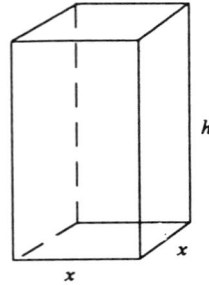

Figure 6.19

(Here we have ignored the material needed to form the flaps).

From equation (1)

$$h = \frac{500}{x^2}$$

Substituting into equation (2) we have

$$A = 2x^2 + \frac{2000}{x}$$

Step 2   Solve the mathematical problem

The amount of material is a minimum when $\dfrac{dA}{dx} = 0$. So differentiating $A(x)$ with respect to x

$$\frac{dA}{dx} = 4x - \frac{2000}{x^2} = 0$$

Solving for x,

$$4x^3 = 2000$$
$$x^3 = 500$$
$$x = \sqrt[3]{500} = 7.94 \, \text{cm}$$

Substituting into equation (1),

$$h = \frac{500}{x^2} = 7.94 \, \text{cm}.$$

The carton with minimum surface area is a cube of size 7.94 cm.

Note in solving realistic problems it is often necessary to make simplifying assumptions so that you can proceed. It is important to list these assumptions carefully.

Exercise 6H

1. A beam has bending moment M (kNm) given by $M = 6x^2 - 12x$ where x is the distance from one end in metres. Find the position of the minimum value of M.

2. A cylindrical fuel storage tank must hold 15 000 litres. Find its dimensions if its surface area is to be minimised. What is the area of metal used in constructing the tank.

3. In an aqueous solution the product xy of the concentrations x and y of OH⁻ and H⁺ ions respectively is a constant at constant temperature. Deduce under what conditions $x + y$ will be a minimum. Find the concentration of H⁺ under minimum conditions at 293°K when $xy = 10^{-4} \, \text{Kmol}^2\text{dm}^{-6}$.

4. For a belt drive the formula relating the power transmitted P to the speed of the belt v is given by

    $$P(v) = Tv - av^3$$

    where T is the tension in the belt and a is a constant.

    (a) Find the speed such that the belt delivers maximum power.
    (b) Sketch the graph of the function $P(v)$.

5. Find the maximum area that can be fenced off from a rectangular field by 80m of fencing using one of the existing walls of the field.

6.   A cylindrical tin can without a lid is made of sheet metal. If A is the surface area of the sheet used and V is the volume of the can, show that

$$V = \tfrac{1}{2}(Ar - \pi r^3)$$

where r is the radius. (Assume that there is no wastage).

If A is given, show that the volume is a maximum when the diameter of the can is twice the height of the can.

7.   A piece of wire forms the circumference of a circle of radius 0.16 m. The wire is cut and bent to form two new circles. Find the radius of each circle so that the sum of the areas of the two circles is a minimum.

8.   A cylinder is such that the sum of its diameter and height is 20 cm. Write the volume (V cm^3) in terms of the radius of its base (r cm). What is the greatest possible value for the volume of the cylinder?

6.5   DIFFERENTIATION OF SPECIAL FUNCTIONS

You know how to differentiate powers of x such as x^2, x^3, x^{-1}, $x^{-\frac{1}{2}}$ etc. But what about the special functions e^x, $\ln(x)$, $\sin(x)$ and $\cos(x)$? The formal definition as a limit would be used to find the general results.

The derivative of the exponential function e^x can be deduced from its introduction in Chapter 2. In TI-92 Activity 2B you saw that the slope of the tangent to e^x at each x value is equal to the function value of e^x at the x value. In symbols we have

$$\frac{d}{dx}e^x = e^x$$

It is this property of e^x that makes it a very important function in mathematics. It is the only function whose derived function is exactly the same form as the function itself.

TI-92 ACTIVITY 6F

The aim of this activity is to investigate the derivatives of the functions e^{ax}, $\ln(ax)$, $\sin(ax)$, $\cos(ax)$ where a is a constant.

(A) Use the **Limit** command in the TI-92 to find the following limits. Compare your answers with the **Differentiation** command.

$$\lim_{h \to 0} \frac{e^{(x+h)} - e^x}{h}$$

$$\lim_{h \to 0} \frac{\ln(x+h) - \ln(x)}{h}$$

$$\lim_{h \to 0} \frac{\sin(x+h) - \sin(x)}{h}$$

$$\lim_{h \to 0} \frac{\cos(x+h) - \cos(x)}{h}$$

(B) Use the TI-92 and the limit definition to investigate the derivative of e^{ax}, $\ln(ax)$, $\sin(ax)$ and $\cos(ax)$ for values of $a = 2, 3, 0.7, 1.3$ and 4.

From the results of your investigation complete the following table.

Function	Derived Function
e^{ax}	
$\ln(ax)$	
$\sin(ax)$	
$\cos(ax)$	

(C) Predict the derivatives of the following functions and check them with the TI-92.

$$e^{3x} + \sin 2x - \cos 4x$$
$$e^{-2x} - \cos 3x + \ln(5x)$$
$$7e^{0.4x} + 0.9 \sin(4x)$$
$$-0.8e^{-1.6x} - 3.8 \cos(2.7x)$$

Attempt each of the following problems 'by hand' and then check your answer using the TI-92.

1. Find the first and second derivatives of each of the following.

    (a) e^{4x}

    (b) e^{-7x}

    (c) $4e^{0.5x}$

    (d) $2e^{-1.3x}$

    (e) $\ln(5x)$

    (f) $3\ln(2x)$

    (g) $\sin(\pi x)$

    (h) $\sin(2x)$

    (i) $4.2\sin(3.1x)$

    (j) $\cos(4x)$

    (k) $\cos(0.2x)$

    (l) $1.5\cos(2\pi x)$

    (m) $0.3e^{0.1x} - 0.7\sin(0.5x)$

    (n) $4\cos 3x - 3\sin 4x$

    (o) $e^{-0.1x} + e^{0.1x}$

    (p) $\ln(2.6x) - 6\ln(0.7x)$

2. Find the equation of the tangent to the graphs of each of the following functions at the given values of x.

    (a) $2\cos 3x$ at $x = 0.5$

    (b) $\ln(x)$ at $x = 1$

    (c) e^x at $x = 2$

    (d) $e^{-0.1x} - 0.6\cos 3x$ at $x = 0$

3. Show that e^{ax} and $\ln(ax)$ have no turning points or points of inflexion.

4. The temperature at a point on a heated rod varies according to the rule

    $$T = 20 + 100e^{-5t}$$

    Find the rate of change of temperature at (a) $t = 0$, and (b) $t = 1$.

5. The population of a yeast culture is modelled by

    $$P(t) = 4.3e^{-2.1t}$$

    Find the rate of change of the population.

6.   A particle moves so that its displacement as a function of time t is modelled by

$$s = 0.3\sin(0.7t)$$

(a)   Find the speed of the particle when $t = 0$ and $t = 1$ seconds.

(b)   Find the acceleration of the particle at these times.

6.6   RULES OF DIFFERENTIATION

Many functions are made up from the basic functions x^n, e^{ax}, $\ln(x)$, $\sin(ax)$ and $\cos(ax)$ by the rules of addition, multiplication, division and composition. For example, the motion of a damped oscillating system can be modelled by

$$x = e^{-2t}\sin(0.6t + 0.7)$$

for appropriate system parameters.   This function is the product of e^{-2t} and the composite function $\sin(0.6t + 0.7)$.

TI-92 ACTIVITY 6G

The aim of this activity is to investigate the derivatives of products and composite functions.

(A)   Use the TI-92 to investigate the derivative of the following products of functions.

(i) $x^2\sin(x)$ (ii) $e^{2x}\cos(3x)$ (iii) $x^3 e^{4x}$

From the results of your investigation suggest a rule for differentiating the product of two functions $f(x) = u(x).v(x)$.   Use your rule to predict the derivatives of the following functions before you do them with the TI-92.

(vi) $x^2\ln(x)$ (vii) $e^x\ln(x)$ (viii) $4\sqrt{x}\sin(5x)$

(B)   Many functions are defined as composite functions of the form $g[h(x)]$, for example, e^{x^2}.

Use the TI-92 to investigate the derivative of the following functions.

(i) e^{x^2} (ii) $\sin(4x^3)$ (iii) $(3-x^2)^8$ (iv) $\cos(\sqrt{x})$

From the results of your investigation suggest a rule for differentiating the composite function $f(x) = g[v(x)]$. Use your rule to predict the derivatives of the following functions before you do them with the TI-92.

(v) $\sin(x^2)$ (vi) $e^{(4x^2-1)}$ (vii) $(1+x)^8$

(viii) $(a+bx)^7$ (ix) $\ln(a+bx)$ where a,b are constants   (x) $\sqrt{1+3x^3}$

To differentiate any function we need a set of rules for differentiating combinations of functions. The set of rules are quoted as follows:

Sum of two functions

If $f(x) = u(x) + v(x)$

then $\dfrac{df}{dx} = \dfrac{du}{dx} + \dfrac{dv}{dx}$

We have already used this rule often in this chapter.

Product of two functions

If $f(x) = u(x).v(x)$

then $\dfrac{df}{dx} = \left(\dfrac{du}{dx}\right).v + u.\left(\dfrac{dv}{dx}\right)$

This is known as the **Product Rule**

Example 6F

Differentiate (a) x^2e^{3x} and (b) $e^{3x}\sin(4x)$.

Solution

The derivative of x^2 is $2x$ and the derivative of e^{3x} is $3e^{3x}$. Applying the Product Rule

$$\frac{d}{dx}x^2 e^{3x} = 2xe^{3x} + 3x^2 e^{3x}$$

The derivative of sin4x is 4cos4x. Apply the Product Rule

$$\frac{d}{dx}e^{3x}\sin(4x) = 3e^{3x}\sin 4x + 4e^{3x}\cos 4x$$

Quotient of two functions

If $f(x) = \dfrac{u(x)}{v(x)}$

then $\dfrac{df}{dx} = \dfrac{v.\dfrac{du}{dx} - u.\dfrac{dv}{dx}}{v^2}.$

This is known as the **Quotient Rule**.

Example 6G

Differentiate $\dfrac{x^2}{\sin 3x}$.

Solution

The derivative of x^2 is 2x and the derivative of sin3x is 3cos3x. Applying the quotient rule

$$\frac{d}{dx}\left(\frac{x^2}{\sin 3x}\right) = \frac{2x.\sin 3x - 3x^2\cos 3x}{(\sin 3x)^2}$$

Composition of two functions

For $f(x) = g[v(x)]$

let $u = v(x)$

so that $f(x) = g(u)$

then $\dfrac{df}{dx} = \dfrac{dg}{du} \cdot \dfrac{du}{dx}.$

This is known as the **Chain Rule**.

░░░░░░░░░░░░░░░ **Example 6H** ░░░░░░░░░░░░░░░░░░░░░░░░░░░░░░░░░░░░░░░

Differentiate $f(x) = \sin(x^2)$.

Solution

Suppose we let $u = x^2$, then the function f can be written as two simple functions

$$f(x) = \sin(u) \text{ and } u = x^2$$

Now $\dfrac{df}{du} = \cos(u)$ and $\dfrac{du}{dx} = 2x.$

Applying the chain rule

$$\frac{df}{dx} = \cos(u).2x = 2x\cos(x^2)$$

░░░░░░░░░░░░░░░ **Example 6I** ░░░░░░░░░░░░░░░░░░░░░░░░░░░░░░░░░░░░░░░

The rate of increase of the radius of a sphere is 0.6 mm per second. Find the rate of increase of the volume of the sphere when the radius is 20 cm.

Solution

If V is the volume of the sphere when its radius is r then

$$V = \frac{4}{3}\pi r^3$$

We need to find the rate of change of volume $\dfrac{dV}{dt}$ given that $\dfrac{dr}{dt} = 0.6 \, \text{mms}^{-1}.$

Apply the chain rule to the equation for V,

$$\frac{dV}{dt} = \frac{dV}{dr} \cdot \frac{dr}{dt} = \frac{4}{3}\pi(3r^2).\frac{dr}{dt} = 4\pi r^2 \frac{dr}{dt}$$

Substituting for $r = 200$ and $\dfrac{dr}{dt} = 0.6$ we have

$$\frac{dV}{dt} = 4\pi(20^2)0.6 = 3016\,\text{mm}^3\text{s}^{-1}$$

Exercise 6J

1.   Apply the rules of differentiation to find the derivatives of the following.

(a) $\sqrt{x}\,e^{2x}$ (b) $x^2 e^{3x}$ (c) $x^5 e^{-2x}$

(d) $x\sin(2x)$ (e) $\sqrt{x}\cos(\pi x)$ (f) $5x^3 e^{3x}$

(g) $\tan(x)$ (h) $x\ln(x)$ (i) $\dfrac{\sin(2x)}{x^2}$

(j) $\dfrac{\sqrt{x}}{e^{3x}}$ (k) $\dfrac{\cos(3x)}{\sin(2x)}$ (l) $\dfrac{e^{2x}+e^{-2x}}{x^2}$

(m) $(3x-1)^5$ (n) $\sqrt{4x+1}$ (o) $\cos(\pi x - 3)$

(p) e^{x^2} (q) $\ln(x^2+1)$ (r) $\ln(3\cos(2x))$

(s) $\sin^2 x + \cos^2 x$ (t) $x^2 e^{-2x}\sin(3x)$ (u) $3\cos(1-4x)$

(v) $\left(6x - \dfrac{1}{x}\right)^3$ (w) $e^{5x}\sin(0.7x)$ (x) $\ln(x^2)$

(y) $\sec x = \dfrac{1}{\cos x}$ (z) $\dfrac{1}{\sqrt{4x+1}}$

Check your answers on the TI-92.

2.   Find the turning points, if any, of the following functions. Graph the functions on the TI-92 to validate your results.

(a) $t^2 + \dfrac{1}{t}$ (b) $\dfrac{(t+2)}{(t-6)}$

(c) $\dfrac{x^2}{(1-x^2)}$ (d) $\dfrac{(x-3)}{(x+1)}$

(e) $e^{-3t}\sin(2t)$ (f) $\dfrac{e^x}{2+3e^x}$

(g) $\dfrac{xe^x}{(x+1)}$ (h) $x + \sin(x)$

3.   A sector is cut from a circular sheet of metal of radius r and bent round to form a cone.

(a)   Show that if φ is the angle of the sector that is removed then

$$\varphi = 2\pi - 2\pi\sin\theta$$

where θ is the semi-vertical angle of the cone.

(b)   Show that the volume of the cone is

$$V = \frac{1}{3}\pi r^3 \sin^2\theta\cos\theta$$

(c)   Find the angle of the sector removed in order that the volume of the cone is a maximum.

4.   The radius r cm of a circular ink blot on a piece of blotting paper t seconds after it was first viewed is given by

$$r(t) = 12 - \frac{9}{t}$$

(a)   Calculate the radius of the blot after 3 seconds.

(b)   Find the time when the blot has radius 3cm.

(c)   Find the rate at which the radius is changing when the radius is 3cm. Is the radius then increasing or decreasing?

(d)   What is the largest value of r?

5.   Find the stationary points of the following functions, and determine their nature. Validate your results by graphing the functions on the TI-92.

(a) $\dfrac{x^2 - 2x + 4}{x^2 + 2x + 4}$

(b) $\sin(x) - \cos(x)$

(c) $x + \sin(x)$

(d) $e^x \cos(x)$

6.   The power delivered into the load X of a class A amplifier of output resistance R is given by

$$P(X) = \frac{V^2 X}{(X + R)^2}$$

where V is the output voltage.

(i)   Find the value of X such that P is a maximum.

(ii)   Sketch a graph of $P(X)$ against X.

7.   Frequency stability in the cathode-coupled oscillator can be studied with the aid of the correction factor

$$f(\alpha) = 1 - \frac{L}{16C}(\alpha - 1/R)^2$$

where L, R and C are constants.

(i)   Show that the maximum value of f is obtained when $\alpha = 1/R$.

(ii)   Sketch a graph of the function $f(\alpha)$.

8.   By considering the derivative of the function

$$f(x) = \sin(x)\tan(x) - 2\ln[\sec(x)]$$

   (a)   show that f steadily increases as x increases from 0 to $\pi/2$,

   (b)   show that the graph has no points of inflexion between these limits.

9.   The mass of gas which will flow through an orifice from pressure p_1 to p_0 is proportional to

$$x^k \sqrt{(1-x^{1-k})}$$

   where k (< 1) is a constant.

   Show that the maximum value of this expression occurs when

$$x = \left(\frac{2k}{1+k}\right)^{1/(1-k)}$$

APPENDIX: PROOFS OF THE RULES OF DIFFERENTIATION

In this section we give formal proofs of the rules for differentiating sums, products, quotients and composition of functions.

1.   **Sum of two functions**.

$$\frac{d}{dx}(u(x) + v(x)) = \lim_{h \to 0} \left\{ \frac{(u(x+h) + v(x+h)) - (u(x) + v(x))}{h} \right\}$$

$$= \lim_{h \to 0} \left\{ \frac{u(x+h) - u(x)}{h} + \frac{v(x+h) - v(x)}{h} \right\}$$

$$= \frac{du}{dx} + \frac{du}{dx}$$

2.   **Product of two functions**.

$$\frac{d}{dx}(u(x)v(x)) = \lim_{h \to 0} \left\{ \frac{u(x+h)v(x+h) - u(x)v(x)}{h} \right\}$$

$$= \lim_{h \to 0} \left\{ \frac{[u(x+h) - u(x)]v(x+h) + u(x)[v(x+h) - v(x)]}{h} \right\}$$

$$= \lim_{h \to 0} \left\{ \frac{[u(x+h) - u(x)]}{h}v(x+h) + u(x)\frac{[v(x+h) - v(x)]}{h} \right\}$$

$$= \frac{du}{dx} \cdot v + u \cdot \frac{dv}{dx} = u'v + uv'$$

To get to the second line we have subtracted $u(x)v(x + h)$ from the first term and added $u(x)v(x + h)$ to the second term in the numerator of the first line.

3.   **Quotient of two functions**.

$$\frac{d}{dx}\left(\frac{u(x)}{v(x)}\right) = \lim_{h\to 0}\left\{\frac{\dfrac{u(x+h)}{v(x+h)} - \dfrac{u(x)}{v(x)}}{h}\right\}$$

$$= \lim_{h\to 0}\left\{\frac{u(x+h)v(x) - u(x)v(x+h)}{hv(x)v(x+h)}\right\}$$

$$= \lim_{h\to 0}\left\{\frac{[u(x+h) - u(x)]v(x) + u(x)[v(x) - v(x+h)]}{hv(x)v(x+h)}\right\}$$

$$= \lim_{h\to 0}\left\{\frac{[u(x+h) - u(x)]}{h}v(x) - u(x)\frac{[v(x+h) - v(x)]}{h}\right\}\frac{1}{v(x)v(x+h)}$$

$$= \frac{\dfrac{du}{dx}\cdot v - u\cdot\dfrac{dv}{dx}}{v^2} = \frac{u'v - uv'}{v^2}$$

4.   **The chain rule for composite functions**.

Start with $f(x) = g[v(x)]$ then

$$\frac{df}{dx} = \lim_{h\to 0}\frac{g[v(x+h)] - g[v(x)]}{h}$$

Let $u = v(x)$ and suppose that $v(x + h) = v(x) + k = u + k$

$$\frac{df}{dx} = \lim_{h\to 0}\frac{g(u+k) - g(u)}{h}$$

$$= \lim_{h\to 0}\left\{\left(\frac{g(u+k) - g(u)}{k}\right)\left(\frac{k}{h}\right)\right\}$$

$$= \lim_{h\to 0}\left\{\left(\frac{g(u+k) - g(u)}{k}\right)\left(\frac{v(x+h) - v(x)}{h}\right)\right\}$$

$$= \frac{dg}{du}\cdot\frac{dv}{dx} = \frac{dg}{du}\cdot\frac{du}{dx}$$

Here we are assuming that as $h \to 0$ then $k \to 0$.

7

Integration

7.1 AREAS AS SUMS

How do you find the area of an irregular shaped region?
How do you find the distance travelled by a space rocket when you know its speed at given time intervals?

These are typical problems which involve summations. Consider the following examples.

Example 7A

Find the area between the function $f(x) = e^x$, and the lines $x = 0$, $x = 1$ and $y = 0$.

Solution

The shaded area in Figure 7.1 shows the area to be found.

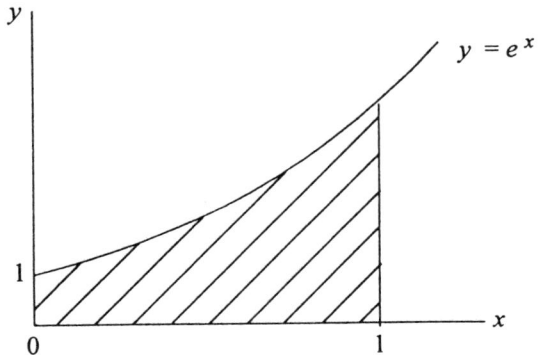

Figure 7.1

The ancient Greeks solved problems of this type and invented a clever method of proceeding. They divided the region into thin strips so that each strip was rectangular in shape. This is shown in Figure 7.2.

The area of each rectangle is easy to calculate and the sum of the areas is then an approximation to the required area. Clearly the thinner the strips, the more accurate is the approximation.

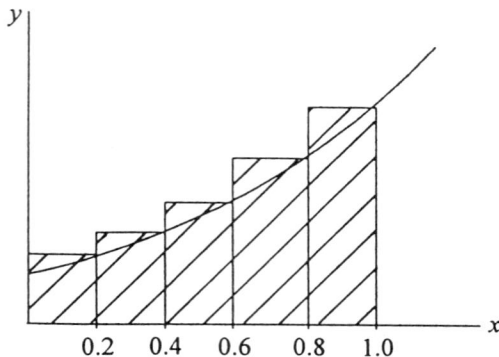

Figure 7.2

Suppose that we take five vertical strips each of equal width 0.2. Then the heights of the five rectangles are $e^{0.2}$, $e^{0.4}$, $e^{0.6}$, $e^{0.8}$ and $e^{1.0}$. The total area of the five rectangles is then

$$\text{Area} = (e^{0.2} \times 0.2) + (e^{0.4} \times 0.2) + (e^{0.6} \times 0.2) + (e^{0.8} \times 0.2) + (e^{1.0} \times 0.2)$$
$$= 1.896$$

The area between $y = e^x$, $y = 0$, $x = 0$ and $x = 1$ is approximately 1.896 (to 4 sf). At this stage we do not know how good is the approximation. By taking double the number of rectangles so that they each have width 0.1 we get the approximation 1.806 (to 4 sf) and with 100 rectangles of width 0.01 we get 1.727. (As we shall show later the actual area is 1.71828 to six significant figures).

Example 7B

A rocket is launched into space. The speed of the rocket at two second time intervals is shown in Table 7.1.

Table 7.1

time (s)	2	4	6	8	10
speed (ms^{-1})	4	16	36	64	100

Estimate the distance travelled during the first ten seconds of the flight.

Solution

In this example we do not know a formula for the speed. Let us assume that during each two second interval the speed of the rocket is constant. The distance travelled during each two second interval is just speed multiplied by time. So we can estimate the distance travelled in ten seconds by summation.

$$\text{Distance travelled} \cong 4 \times 2 + 16 \times 2 + 36 \times 2 + 64 \times 2 + 100 \times 2$$
$$= 440\,\text{m}.$$

This is clearly a simplification based on the assumption of constant speeds. We could improve the approximation by assuming a linear speed in each interval. These two cases are shown in Figure 7.3.

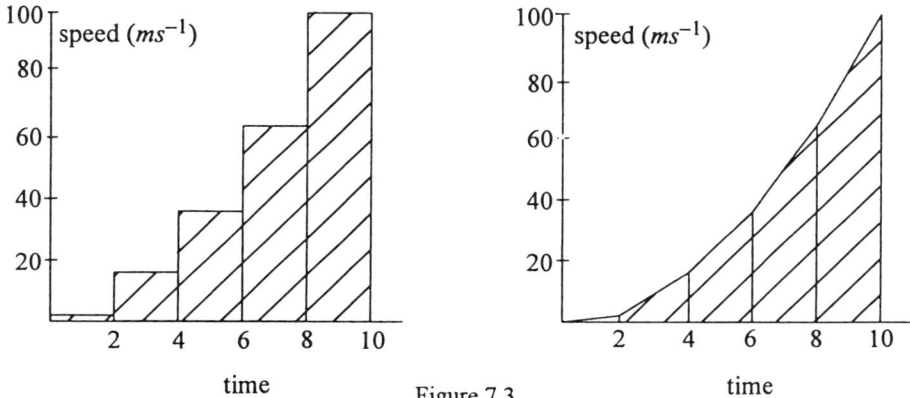

Figure 7.3

Again the summation is an approximation to the area of the region.

Exercise 7A

1.   Estimate the area of the following regions, choosing ten sub-intervals in each case.

(a)   Between $y = x^2$, $y = 0$, $x = 0$ and $x = 3$.

(b)   Between $y = x^3$, $y = 0$, $x = 1$ and $x = 2$.

(c)   Between $y = \sin x$, $y = 0$, $x = 0$ and $x = \pi/3$.

(d)   Between $y = e^{-2x}$, $y = 0$, $x = -1$ and $x = 3$.

2.   A particle moves so that its velocity (in ms^{-1}) at time t is given by

$$v = \frac{t(8 - t^3)}{4} \qquad 0 \le t \le 2.$$

(a)   Sketch a graph of v against t.

(b)   Estimate the distance travelled during the first two seconds of the motion of the particle.

3.   On a car journey the speed of a car is recorded every 5 minutes and is shown in the following table. The car starts from rest at $t = 0$.

Table 7.2

time (in min)	5	10	15	20	25	30	35	40
speed (in mph)	20	30	30	15	20	25	15	0

(a)   Estimate the distance travelled during the journey.

(b)   How could we obtain a better estimate of the distance travelled?

7.2   AREAS AS INTEGRALS

In this section we explore the process of approximating areas by taking thinner and thinner strips and more of them. This leads to an important topic in calculus called **integration**.

Consider the problem of finding the area between the graph of $y = x^2$ and the lines $x = 0$ and $x = 2$ (see Figure 7.4).

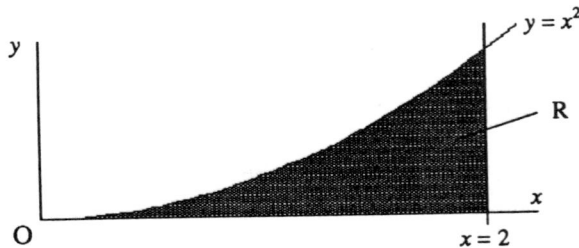

Figure 7.4

Suppose that we divide the region into 5 thin strips each of width 0.4. The area of the region R can then be approximated by the following sum which we denote by

$$A_5 = 0.4(0.4^2) + 0.4(0.8)^2 + 0.4(1.2)^2 + 0.4(1.6)^2 + 0.4(2)^2$$

$$= 3.52$$

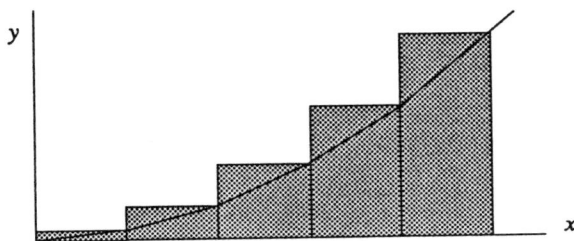

Figure 7.5

If we double the number of strips and halve their width we obtain the approximation

$$A_{10} = 0.2(0.2)^2 + 0.2(0.4)^2 + 0.2(0.6)^2 + 0.2(0.8)^2 + 0.2(1.0)^2 + 0.2(1.2)^2$$

$$+ 0.2(1.4)^2 + 0.2(1.6)^2 + 0.2(1.8)^2 + 0.2(2.0)^2$$

$$= 3.08$$

Table 7.3 shows what happens as we keep doubling the number of strips.

Table 7.3

number of strips, n	strip width	A_n
5	0.4	3.52
10	0.2	3.08
20	0.1	2.87
40	0.05	2.7675

In this activity we have an example of a limiting process. As A_n increases we can show that the sum approaches the value $2.\dot{6}6$.

TI-92 ACTIVITY 7A

Figure 7.6 shows a simple program for automating this limiting process. Enter this program into your calculator.

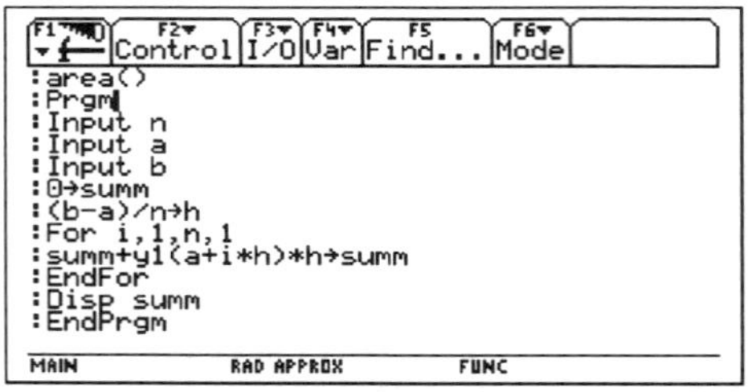

Figure 7.6   Program for finding the area under a graph using rectangles

(A)   Use the program to reproduce the results in Table 7.3. How many strips are needed to give two equal answers correct to six decimal places?

(B)   Use the program to find the following areas correct to 2 significant figures. (Remember to edit the function $y1(x)$.)

    (a)   Between $y = x^2$, $y = 0$, $x = 0$ and $x = 3$.
    (b)   Between $y = x^3$, $y = 0$, $x = 1$ and $x = 2$.
    (c)   Between $y = e^x$, $y = 0$, $x = 0$ and $x = 2$.
    (d)   Between $y = e^{-2x}$, $y = 0$, $x = -1$ and $x = 3$.

Table 7.4 shows the results of part (d) starting with $n = 4$. To two significant figures we deduce that the area is 3.7. In a later section we shall be able to evaluate the area exactly. To six significant figures the area is 3.69329. So this method of rectangular strips converges very slowly.

Table 7.4

number of strips	strip width	A_n
4	1	1.15613
8	2^{-1}	2.14941
16	2^{-2}	2.84659
32	2^{-3}	3.25084
64	2^{-4}	3.46727
128	2^{-5}	3.57908
256	2^{-6}	3.63588
512	2^{-7}	3.66451
1024	2^{-8}	3.67888

From this activity you can deduce that for these functions

$$\text{area of region } R = \lim_{n \to \infty} A_n$$

In fact this result will be true for any continuous function.

More generally consider the problem of finding the area between the graph of $y = f(x)$, $x = a$, $x = b$ and $y = 0$ (see Figure 7.7).

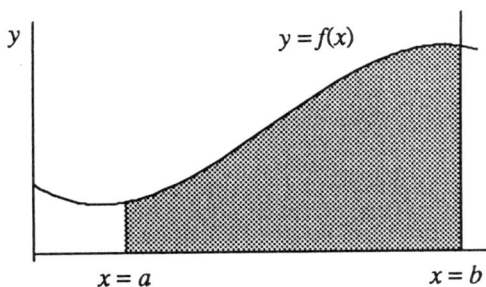
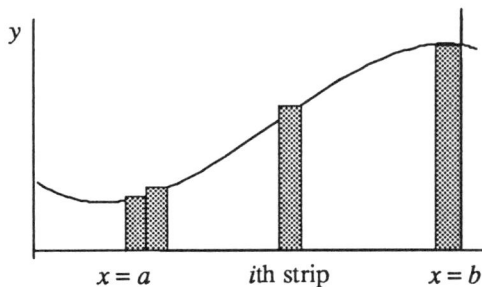

Figure 7.7   Figure 7.8

Suppose that we divide the region into n vertical strips each of the same width $h = (b - a)/n$ (see Figure 7.8). The area of the ith rectangular strip is given by the expression

$$S_i = f(x_i)h$$

where $x_i = a + ih$. The sum of the areas of the n rectangles is then

$$A_n = \sum_{i=1}^{n} f(x_i)h = \sum_{i=1}^{n} f(a+ih)h$$

which is an approximation to the area of the region required.

If we increase the number of rectangular strips, and hence make them thinner, the value of A_n becomes closer to the value of the required area.

In the limit, as $n \to \infty$, the summation should equal the area. Formally we write

$$\text{Area} = \lim_{n \to \infty} A_n = \lim_{n \to \infty} \sum_{i=1}^{n} f(a+ih)h \qquad \text{with } h = \frac{b-a}{n}.$$

The limiting process is a complicated one but for "well behaved continuous functions" it does converge. Finding the limit of a sum is called **integration** and in the above case the **integral of the function** f is an area. We use a special symbol $\int$ to denote it. We write

$$\int_a^b f(x)dx = \lim_{n \to \infty} \sum_{i=1}^{n} f(a+ih)h \qquad \text{with } h = \frac{b-a}{n}$$

For example, the area in Example 7A is written formally as $\int_0^1 e^x \, dx$ and the distance travelled by the rocket in Example 7B can be written as $\int_0^{10} v \, dt$ where $v(t)$ is the velocity function.

For two simple functions $f(x) = c$ and $f(x) = mx$ where m and c are positive constants, we can evaluate the integral of f by finding the area summations exactly. For other functions we need a set of standard rules.

▓▓▓▓▓▓▓▓ **Example 7C** ▓▓▓▓▓▓▓▓

Find the integrals of the functions $f(x) = c$ and $f(x) = mx$ between $x = a$ and $x = b$.

Solution

The values of the integrals are shown by the areas in Figure 7.9.

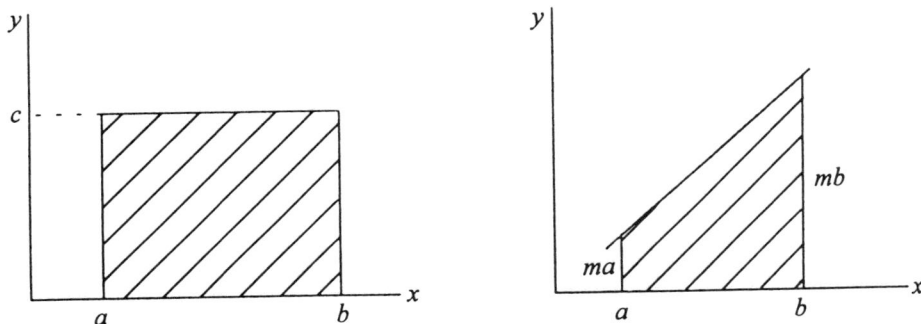

Figure 7.9

(a) $f(x) = c$.

The region is a rectangle whose area is $(b-a)c$; so we write

$$\int_a^b c\,dx = (b-a)c$$

(b) $f(x) = mx$.

The region is a trapezium whose area is $\frac{1}{2}(b-a)[ma+mb]$; so we can write

$$\int_a^b mx\,dx = (b-a)[\tfrac{1}{2}m(b+a)] = \tfrac{1}{2}m(b^2 - a^2)$$

TI-92 ACTIVITY 7B

The aim of this investigation is to use the TI-92 to evaluate the summation limit and show how the integral of a function leads to a new function. We begin with $a = 0$. Clear the HOME screen and select AUTO mode.

(A)   Enter the expressions $f(ih)h$ and form the sum and limit so that you obtain the home screen shown in Figure 7.10.

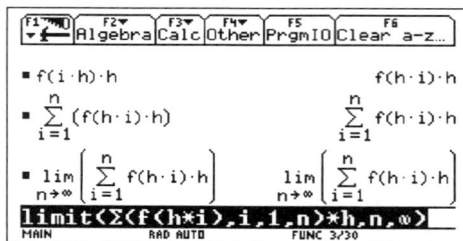

Figure 7.10   Forming the infinite summation

(B)   Now we will use the infinite summation process to evaluate the limit to find the integral

$$\int_0^b x \, dx$$

First define $f(x) = x$ (using **F4** and select 1:Define).   Then highlight the expression containing limit and press **ENTER**.   Set $h = b/n$ and press **ENTER**. Figure 7.11 show the home screen for this activity.

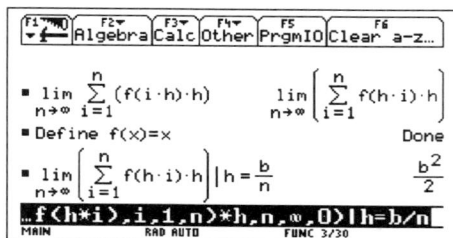

Figure 7.11   Finding the limit of a summation to evaluate $\int_0^b x \, dx$

This activity shows that

$$\int_0^b x\,dx = \tfrac{1}{2}b^2$$

which is the same result as in Example 7C (b) with $a = 0$ and $m = 1$.

(C)   Repeat activity (B) for $f(x) = x^2$ and show that

$$\int_0^b x^2\,dx = \tfrac{1}{3}b^3$$

Deduce the value of $\int_0^2 x^2\,dx$ and hence confirm the result found in TI-92 Activity 7A.

(D)   Repeat the process for the following functions keeping a record of your results:

$$x^3, x^4, x^5, x^6.$$

Can you see a pattern emerging?
What can you conclude is the general result of evaluating

$$\int_0^b x^m\,dx = ?$$

(E)   Now investigate the integral of the following functions between $x = 0$ and $x = b$

$$4x^2,\ 5x^5,\ 3x^4,\ x^2 + x^3,\ x + x^4$$

Deduce rules for the following expressions:

$$\int_0^b rx^m\,dx \qquad\qquad \text{where } r \text{ is a constant}$$

$$\int_0^b x^p + x^m\,dx$$

Summary

The results of TI-92 Activity 7B can now be summarised to provide some basic rules for integration.

$$\int_0^b x^m \, dx = \frac{b^{m+1}}{m+1}$$ where m is a positive integer

$$\int_0^b rx^m \, dx = r\int_0^b x^m \, dx$$ provided r is a constant

$$\int_0^b x^p + x^m \, dx = \int_0^b x^p \, dx + \int_0^b x^m \, dx$$

i.e. the integral of a sum of functions is the sum of the integrals.

Notation

The process of finding areas under curves is called **definite integration**. It is a special summation process. If we are finding the area between $y = f(x)$, $y = 0$, $x = a$ and $x = b$ then we write the definite integral as

$$\int_a^b f(x)dx$$ (1)

where a and b are called **the limits** of the integration. Expression (1) has a definite value. Figure 7.12 shows the areas.

$$\int_a^b f(x)dx, \quad \int_0^a f(x)dx, \quad \int_0^b f(x)dx$$

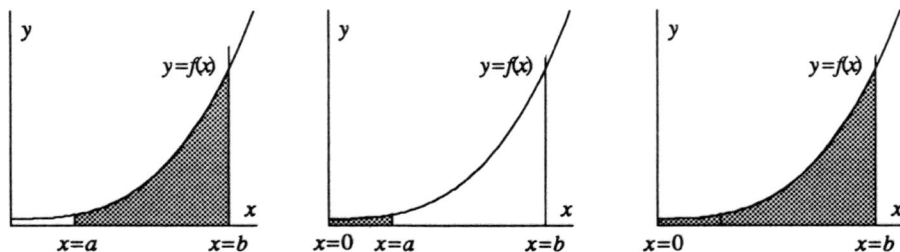

Figure 7.12

From the figures we can deduce that

$$\int_a^b f(x)dx = \int_0^b f(x)dx - \int_0^a f(x)dx$$

For the special case, $f(x) = x^n$ we can write

$$\int_a^b x^n\, dx = \int_0^b x^n\, dx - \int_0^a x^n\, dx$$
$$= \frac{b^{n+1}}{n+1} - \frac{a^{n+1}}{n+1} \qquad\qquad (2)$$

We shall see that integration is more than just an area under a graph. It is a process in its own right that leads to a 'new' function. The process is very useful in mathematical problem solving.

Starting with x^n a rule for forming the right hand side of expression (2) is "add 1 to n and divide by $n + 1$". This leads to the important rule for integrating the function x^n

---

if n is a positive integer then

$$\text{integral of } x^n \Rightarrow \frac{x^{n+1}}{n+1}$$

The 'new' function $\dfrac{x^{n+1}}{n+1}$ is called the **indefinite integral** of x^n. Formally we write the indefinite integral using the symbol $\int$ without any limits. (In Section 7.5 we see that we should add a constant, called the constant of integration to this new function.)

Example 7D

Find the indefinite integral of the functions (a) x^4, and (b) x^7.

Solution

(a) $\displaystyle\int x^4 dx = \frac{x^5}{5}$

(b) $\displaystyle\int x^7 dx = \frac{x^8}{8}$

We can use the indefinite integral to evaluate a definite integral. Consider equation (2).

$$\frac{b^{n+1}}{n+1} = \text{value of } \frac{x^{n+1}}{n+1} \text{ when } x = b;$$

$$\frac{a^{n+1}}{n+1} = \text{value of } \frac{x^{n+1}}{n+1} \text{ when } x = a$$

and $\displaystyle\int_a^b x^n\, dx$ is the difference between these values.

The notation used to show these ideas is

$$\int_a^b x^n\, dx = \left.\frac{x^{n+1}}{n+1}\right|_a^b \quad\begin{array}{l}\leftarrow \text{upper limit}\\[6pt]\leftarrow \text{lower limit}\end{array}$$

Example 7E

Use the indefinite integrals of Example 7D to evaluate

(a) $\displaystyle\int_1^3 x^4\, dx$ and   (b) $\displaystyle\int_{-1}^1 x^7\, dx$.

Solution

(a) $\int_1^3 x^4 \, dx = \dfrac{x^5}{5}\bigg|_1^3 = \dfrac{3^5}{5} - \dfrac{1^5}{5} = \dfrac{242}{5} = 48\tfrac{2}{5}$

(b) $\int_{-1}^1 x^7 \, dx = \dfrac{x^8}{8}\bigg|_{-1}^1 = \dfrac{(1)^8}{8} - \dfrac{(-1)^8}{8} = 0$

Exercise 7B

1.   Find the values of the following integrals:

(a) $\int_0^3 x^4 \, dx$

(b) $\int_0^1 x^5 \, dx$

(c) $\int_0^{1.2} x^3 \, dx$

(d) $\int_1^3 x^4 \, dx$

(e) $\int_2^4 3x^5 \, dx$

(f) $\int_{-1}^2 4x^2 \, dx$

(g) $\int_1^2 x - x^2 \, dx$

(h) $\int_{-2}^1 0.1x^3 - \dfrac{x^4}{2} \, dx$

(i) $\int_0^4 \dfrac{x^3}{7} + \dfrac{x}{11} - 2 \, dx$

(j) $\int_{-0.5}^{1.5} 5x^2 + 4x^5 \, dx$

2.   Write each of the following areas as an integral and evaluate them.

(a)   Between $y = 2x^2$, $y = 0$, $x = 1$ and $x = 4$.

(b)   Between $y = x - x^4$, $y = 0$, $x = 0$ and $x = 1$.

(c)   Between $y = 2 - x^2$, $y = 0$, $x = -1$ and $x = 1$.

3.   A particle moves so that its velocity (in ms^{-1}) at time t is given by

$$v = \frac{t(8 - t^3)}{4}$$

(a)   Sketch a graph of v against t.

(b)   Find the distance travelled during the first 2 seconds of the motion of the particle.

4.   The work done by an object when moving in a straight line a distance d against a constant force F is defined by $W = Fd$.
An object moves along the x-axis between $x = 1$ and $x = 4$ against a variable force $F(x) = 20 - x^2$.

(a)   Sketch a graph of $F(x)$ against x and show the value of the work done.

(b)   Evaluate the work done by the object.

7.3   INTEGRATION

In TI-92 Activity 7B we evaluated the summation limit between 0 and b. From the results we deduced that

$$\int_a^b x^m \, dx = \frac{b^{m+1}}{(m+1)} - \frac{a^{m+1}}{(m+1)} \qquad (7.1)$$

for integer m.
This is the limit of the sum of the rectangular areas shown in Figure 7.13.

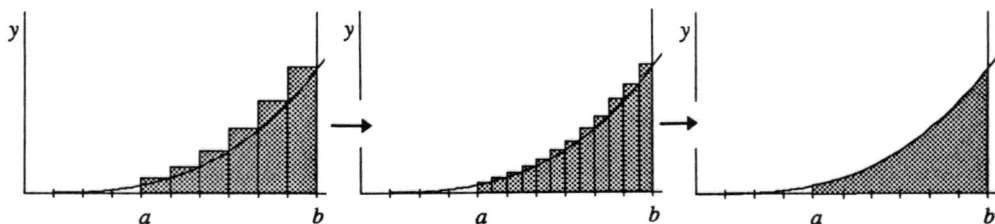

Figure 7.13   Summing between a and b

In symbols we write

$$\int_a^b x^m \, dx = \lim_{n \to \infty} \sum_{i=1}^{n} f(a+ih)h$$

with $h = \dfrac{b-a}{n}$ and $f(x) = x^m$.

TI-92 ACTIVITY 7C

The aim of this Activity is to investigate the integral of the exponential function $f(x) = e^{\alpha x}$ Clear the HOME screen and the Y = Editor.

(A)   Enter the expressions $f(a + ih)h$ and form the sum and limit so that you obtain the home screen shown in Figure 7.14.

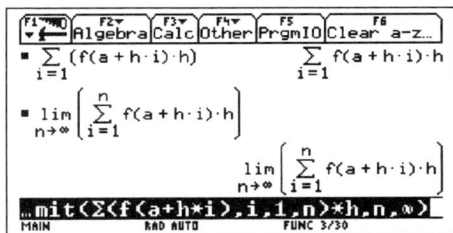   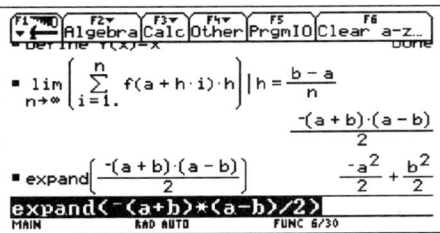

Figure 7.14   Figure 7.15

(B)   Define $f(x) = x$ (using F4) and evaluate the limit to find the integral

$$\int_a^b x \, dx$$

Use **F2** and select 3:expand(to show that the answer is $\dfrac{b^2}{2} - \dfrac{a^2}{2}$. Figure 7.15 shows the HOME screen dump for this activity.

Repeat for x^2 and x^3 and check that you obtain the formula in equation (7.1).

(C)   Define $f(x) = e^x$ and evaluate the limit to find the integral

$$\int_a^b e^x \, dx$$

(D)   Repeat the process for the following functions keeping a record of your results

$$e^{2x}, e^{3x}, e^{4x}, e^{5x}$$

Can you see a pattern emerging?
What can you conclude is the general result of evaluating

$$\int_a^b e^{cx} dx = ?$$

where c is a constant.

Deduce the formula for the indefinite integral

$$\int e^{cx} dx =$$

where c is a constant.

Summary

The results of TI-92 Activity 7C show that

$$\int e^{cx} dx = \frac{1}{c} e^{cx}$$

where c is a constant.

Example 7F

Find the area between the function $y = e^{-2x}$, the x-axis, $x = -1$ and $x = 3$.

Solution

$$\text{Area} = \int_{-1}^3 e^{-2x} dx$$

$$= -\frac{1}{2} e^{-2x} \Big|_{-1}^3$$

$$= -\frac{1}{2} e^{-6} + \frac{1}{2} e^2$$

$$= 3.69329$$

This agrees to 2 significant figures with the area found by summing 1024 rectangular strips in TI-92 Activity 7A part (B) (d) shown in Table 7.4.

Exercise 7C

1. Find the values of the following integrals:

   (a) $\displaystyle\int_0^1 e^x \, dx$ (b) $\displaystyle\int_1^2 e^{3x} \, dx$ (c) $\displaystyle\int_{-1}^1 e^{7x} \, dx$

   (d) $\displaystyle\int_0^2 x^2 + e^{2x} \, dx$ (e) $\displaystyle\int_1^2 x^4 - 3e^{4x} \, dx$

   (f) $\displaystyle\int_{-1}^1 0.1x^3 + 2e^x \, dx$ (g) $\displaystyle\int_0^2 e^x - e^{2x} + x \, dx$

2. Write each of the following areas as an integral and evaluate them.

   (a) Between $y = e^x$, $y = 0$, $x = 0$ and $x = 2$.

   (b) Between $y = e^{2x}$, $y = 0$, $x = -1$ and $x = 3$.

   (c) Between $y = e^{3x} - 1$, $y = 0$, $x = 0$ and $x = 1.5$.

7.4   THE FUNDAMENTAL THEOREM OF CALCULUS

The integral of a function is defined as the limit of a summation process

$$\int_a^b f(x)dx = \lim_{n \to \infty} \sum_{i=1}^n f(a+ih)h \qquad \text{with } h = \frac{b-a}{n}$$

An application of this definition is the area of the region between $y = f(x)$, $x = a$, $x = b$ and $y = 0$. From this we defined the indefinite integrals

$$\int x^m \, dx = \frac{x^{m+1}}{m+1}$$

$$\int e^{cx} \, dx = \frac{1}{c}e^{cx}$$

The Fundamental Theorem of Calculus provides a link between integration and differentiation. Consider the derivatives of the formulas above

$$\frac{d}{dx}\left(\frac{x^{m+1}}{m+1}\right) = (m+1)\frac{x^m}{(m+1)} = x^m$$

$$\frac{d}{dx}\left(\frac{1}{c}e^{cx}\right) = \frac{1}{c}(ce^{cx}) = e^{cx}$$

Integration is closely related to differentiation and we often think of them as opposite processes (see Figure 7.16).

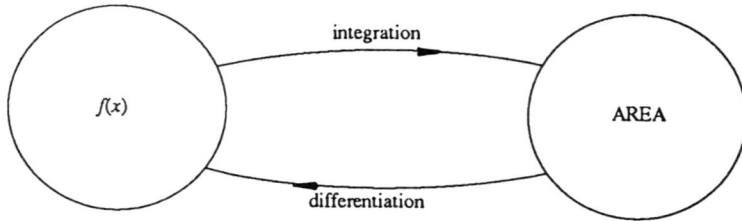

Figure 7.16

It is essential to appreciate the significance of the Fundamental Theorem which is true for all our basic functions. For example, consider the problem of evaluating the integral of cos x. We could try to evaluate the limit of the summation. However, it is easier to use our knowledge of differentiation. Since

$$\frac{d}{dx}\sin x = \cos x$$

we can deduce from the Fundamental Theorem of Calculus

$$\int \cos x \, dx = \sin x$$

Now because the derivative of a constant is zero, we could also write

$$\frac{d}{dx}(\sin x + c) = \cos x \qquad\qquad \text{for any constant } c.$$

So $$\int \cos x \, dx = \sin x + c$$

When evaluating indefinite integrals it is **essential** to add the constant c which is called the **constant of integration**.

Summary

The results of Sections 7.1-7.5 can now be brought together into the following table of integrals.

Table 7.5

Function f	Integral of $f = g(x)$
x^n $(n \neq -1)$	$\dfrac{x^{n+1}}{(n+1)} + c$
$\dfrac{1}{x}$	$\ln(x) + c$
e^{ax}	$\dfrac{1}{a}e^{ax} + c$
$\sin(ax)$	$-\dfrac{1}{a}\cos(ax) + c$
$\cos(ax)$	$\dfrac{1}{a}\sin(ax) + c$

In each case c is an unknown constant called a **constant of integration**.
This table can be used to evaluate definite integrals

$$\int_a^b f(x)\, dx = g(b) - g(a)$$

We often use the notation $\int_a^b f(x)\, dx = \left[g(x)\right]_a^b = g(b) - g(a)$.

The following example shows the method of approach when evaluating integrals using Table 7.5.

░░░░░░░░░░ **Example 7G** ░░░░░░░░░░░░░░░░░░░░░░░░░░░░░░░░░

Evaluate $\displaystyle\int_0^{\pi/4} \cos 2x\, dx$.

Solution

From the table of standard integrals (Table 7.5)

$$\int \cos 2x\, dx = \frac{1}{2}\sin 2x + c$$

Hence

$$\int_0^{\pi/4} \cos 2x \; dx = \left[\frac{1}{2} \sin 2x + c \right]_0^{\pi/4}$$

$$= \left(\frac{1}{2} \sin \frac{\pi}{2} + c \right) - \left(\frac{1}{2} \sin 0 + c \right)$$

$$= \frac{1}{2}$$

Notice that the constant of integration cancels. In practice when evaluating definite integrals we ignore c.

Example 7H

Evaluate $\int_2^4 \frac{1}{x} \, dx$.

Solution

From the table of standard integrals (Table 7.5)

$$\int \frac{1}{x} \, dx = \ln(x) + c$$

so $\int_2^4 \frac{1}{x} \, dx = \left[\ln(x) \right]_2^4 = \ln(4) - \ln(2) = \ln\left(\frac{4}{2} \right) = \ln 2$

The TI-92 can be used to evaluate integrals. The following commands will evaluate the integral in Example 7G. Choose 2:∫(integrate from the **F3** menu or use the ∫ from **2ⁿᵈ 7**.

Type cos 2x, x, 0, π/4) and **ENTER**.

Figure 7.17 shows the home screen for this activity and for Example 7H. Figure 7.18 shows the home screen for some indefinite integrals. Notice that the TI-92 shows the formal expression of the definite integral, but omits the constant of integration for indefinite integrals.

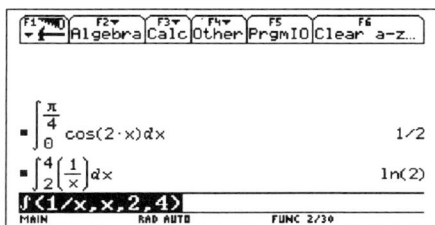

Figure 7.17 Definite integrals

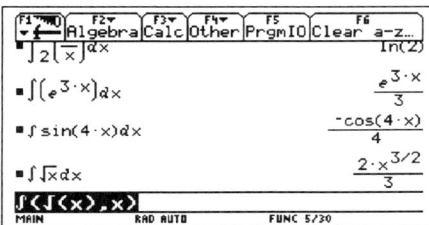

Figure 7.18 Indefinite integrals

Example 7I

The gradient of a curve which passes through the point $(-1,2)$ is given by $3 - x^2$. Find the equation of the curve.

Solution

Suppose that the equation of the curve is $y = f(x)$. Then the gradient is the slope of the tangent $f'(x)$ which is given by $3 - x^2$. Hence

$$f'(x) = 3 - x^2$$

Integrating

$$f(x) = \int (3 - x^2)\,dx$$

$$= 3x - \frac{x^3}{3} + c$$

The point $(-1,2)$ lies on the curve so $f(-1) = 2$. Using the formula above

$$f(-1) = -3 + \frac{1}{3} + c = -\frac{8}{3} + c$$

but $f(-1) = 2$ so that $c = 2 + \frac{8}{3} = \frac{14}{3}$. The equation of the curve is

$$y = 3x - \frac{x^3}{3} + \frac{14}{3}$$

████████████ **Exercise 7D** ████████████

In each problem evaluate the integrals 'by hand' and then check your answers using the TI-92.

1. Find the indefinite integrals of the following functions.

(a) $4x^3$

(b) $3x^5$

(c) $x^{\frac{1}{2}}$

(d) $13x^2 - 7x^3$

(e) $3x^5 + 2x^3 - x + 4$

(f) $6 + 3x - 2x^2$

(g) $(4x+2)^2$

(h) $(1-x)^2$

(i) $\dfrac{1}{x^2}$

(j) $x^{-0.7}$

(k) $1.7x^{-2.3}$

(l) $3x^{-1}$

(m) $5x^{-6} + 3x^{-2}$

(n) $2x^{0.3} + \dfrac{1}{x}$

(o) $9x^{17} - 2x^4 + 3x^{-2}$

2. Find the indefinite integrals of the following functions.

(a) e^{2x}

(b) e^{-5x}

(c) $e^{0.1x}$

(d) $3e^{4x}$

(e) $6e^{6x}$

(f) $-0.9e^{-0.5x}$

(g) $4e^{3x} - 3e^{-2x}$

(h) $0.6e^{3.1x} - 0.9e^{-0.3x}$

3. Find the indefinite integrals of the following functions.

(a) $\sin(5x)$

(b) $\cos(1.5x)$

(c) $4\sin(3x)$

(d) $3\sin(2x) - 2\cos(3x)$

(e) $2\sin(\pi x)$

(f) $1.5\cos(3\pi x)$

(g) $2\sin(wx)$ where w is a constant

(h) $1.5\cos(7x) + 0.3\sin(2x)$

(i) $e^{0.1x} + 2\sin(\pi x)$

(j) $3x^2 + 4.2e^{-0.6x} + \cos(0.9x)$

(k) $\dfrac{1}{3x} + 0.5\sin(5x)$

4. Evaluate the following integrals.

(a) $\int_1^2 x^2 dx$ (b) $\int_0^1 3x + 6dx$ (c) $\int_{-1}^1 x^2 - 3x + 4dx$

(d) $\int_0^1 2e^{3x} dx$ (e) $\int_1^3 e^x - e^{-2x} dx$ (f) $\int_{-1}^1 5e^{0.2x} dx$

(g) $\int_0^\pi \sin x dx$ (h) $\int_0^{\pi/2} \cos x dx$ (i) $\int_0^4 7e^{-0.1x} - 2x^{\frac{1}{2}} dx$

5. Use integration to evaluate the area of the following regions:

(a) between $y = 1 - x^2 + x^4$, $y = 0$, $x = 0$ and $x = 1$,

(b) between $y = \sin x$ and the x-axis between $x = 0$ and $x = \pi$,

(c) between $y = e^{-2x}$, $y = 0$, $x = 1$ and $x = 3$,

(d) between $y = 2e^x$, $y = 1$, $x = 0$ and $x = 2$.

6. Integrate the following.

(a) $(x-1)(x-2)$ (b) $x^2(x+2)$ (c) $\dfrac{1+x}{x^3}$

(d) $\dfrac{x^4+1}{x^2}$ (e) $\dfrac{a}{x^2}+b$ where a, b are constants

(f) $ax^2 + bx + c$ where a, b and c are constants.

7. The gradient of a curve which passes through the point $(1,1)$ is given by $2 + 2x - x^2$. Find the equation of the curve.

8. The gradient of a curve which passes through the point $(0,1)$ is given by $1 + x^2$. Find the equation of the curve.

9.   Evaluate the following integrals.

(a) $\int t^2 dt$ (b) $\int 3t + 1 dt$ (c) $\int t^3 + 2t^2 dt$

(d) $\int \frac{1}{t^2} dt$ (e) $\int 5t^7 + 4t^3 - t^{-1} dt$ (f) $\int (t-1)^2 dt$

(g) $\int at^2 + bt + c \, dt$ where a, b, c are constants

(h) $\int e^{2t} dt$ (i) $\int 2e^{0.1t} dt$ (j) $\int 2\sin(\pi t) dt$

(k) $\int \frac{1}{t} dt$ (l) $\int 5\cos(0.1t) dt$ (m) $\int \sin(t) + \cos(t) dt$

(n) $\int \frac{1}{v^2} dv$ (o) $\int p^{-\frac{1}{2}} dp$ (p) $\int u^2 + 3u + 8 du$

(q) $\int \frac{1}{w} dw$ (r) $\int e^{2u} du$ (s) $\int \sqrt{y} dy$

10.   The acceleration of an object is related to the velocity through the equation $a = \dfrac{dv}{dt}$.

(a)   If $a = t^2$ find v, given that $v = 2$ when $t = 0$.

(b)   If $a = t + 1$ find v, given that $v = 1$ when $t = 1$.

(c)   If $a = 3\cos(t)$ find v.

11.   The force on an object is related to the potential energy through the equation $F = \dfrac{dV}{dx}$.

(a)   Find the potential energy due to gravity if $F = mg$.

(b)   Find the potential energy in an elastic string if $F = k(x-L)$.

7.5   MORE ABOUT AREAS

The introduction to integration has linked definite integrals to areas. However you must be careful in this interpretation.   Consider the value of the following integral.

$$\int_{-1}^{1} x\,dx = \left[\frac{x^2}{2}\right]_{-1}^{1} = \frac{1}{2} - \frac{1}{2} = 0$$

Does this mean that the area between the function $f(x) = x$ and the x-axis between $x = -1$ and $x = 1$ is zero. Figure 7.19 shows this area.

Clearly the area of triangle $A = \frac{1}{2}$ and the area of triangle $B = \frac{1}{2}$ so the total shaded area $= 1$. But the value of the integral is 0.

To understand this anomaly consider the two integrals $\int_{-1}^{0} x\,dx$ and $\int_{0}^{1} x\,dx$.

Figure 7.19

$$\int_{-1}^{0} x\,dx = \left[\frac{1}{2}x^2\right]_{-1}^{0} = 0 - \frac{1}{2} = -\frac{1}{2}$$

$$\int_{0}^{1} x\,dx = \left[\frac{1}{2}x^2\right]_{0}^{1} = \frac{1}{2} - 0 = \frac{1}{2}$$

When the function values are negative the integral of the function will be negative. Since areas are positive we can interpret the integral as an area by ignoring the sign of the integral. Hence

$$\text{Area } A = \left|\int_{-1}^{0} x\,dx\right| = \left|-\frac{1}{2}\right| = \frac{1}{2}$$

It is important to interpret the value of an integral as an area with caution. It is advisable to sketch the graph of the function to be integrated and then integrate the function in portions to take account of the signs. Figure 7.20 shows the HOME screen and the GRAPH screen for finding the area between $y = x^2 - 4$, $x = 0$, $x = 3$ and $y = 0$.

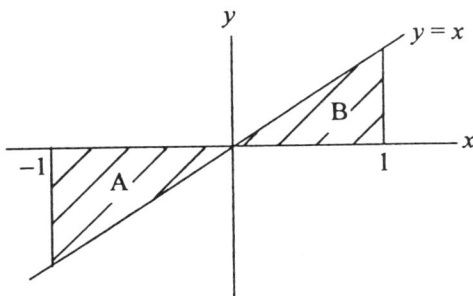

In the HOME screen we have $\int_0^2 x^2 - 4x = -16/3$. (Its value is negative because the graph is below the x-axis.) The required area is given by

$$\left(\int_2^3 x^2 - 4dx\right) - \left(\int_0^2 x^2 - 4dx\right) = 23/3 = 7.6\dot{6}$$

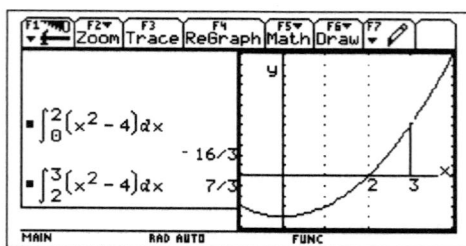

Figure 7.20

Exercise 7E

1. Sketch the graph of $y = x^5$. Show that

   $$\int_{-2}^2 x^5 dx = 0$$

   Use your sketch to explain your answer.

2. Calculate the shaded areas in the following diagrams.

   (a)

   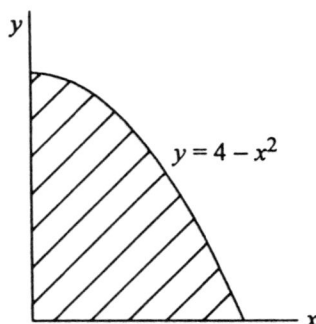

   $y = 4 - x^2$

   (b)

   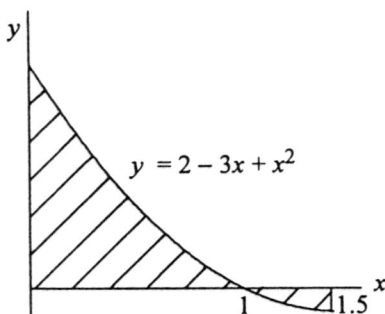

   $y = 2 - 3x + x^2$

(c)   (d)

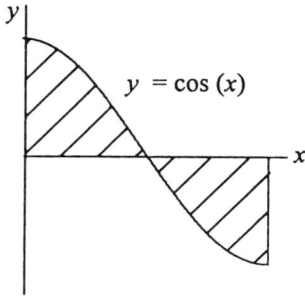

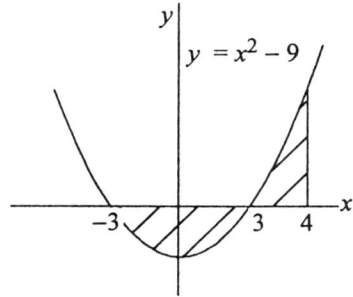

Figure 7.21

3.   Evaluate the following definite integrals and interpret the values in terms of areas.

(a) $\int_{-1}^{1} x - 2\,dx$ (b) $\int_{0.5}^{2} \frac{1}{x}\,dx$

(c) $\int_{0}^{2} x^2 - x\,dx$ (d) $\int_{-1}^{1} 2e^{-x}\,dx$

(e) $\int_{-\pi/2}^{\pi/2} \sin(x)\,dx$ (f) $\int_{0}^{1} 3e^{-x} - 2\,dx$

4.   In the following diagrams the graphs are symmetrical about either the x or y axes. The area of part of the region is given. Deduce the area of the whole shaded region and the value of the given integral.

(a)

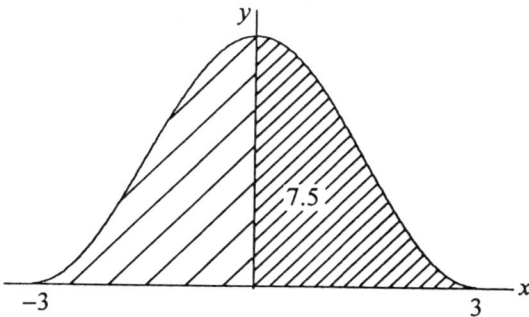

Evaluate $\int_{-3}^{3} f(x)\,dx$.

(b)

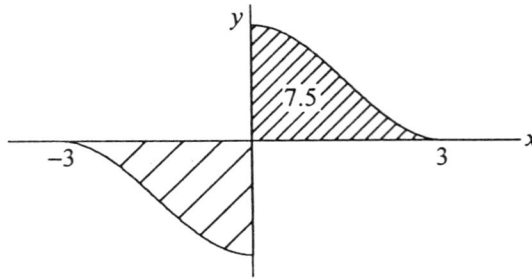

Evaluate $\int_{-3}^{3} f(x)\,dx$.

(c)

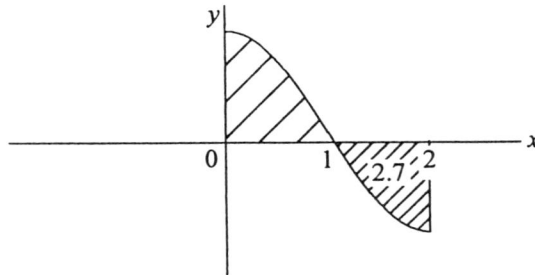

Evaluate $\int_{0}^{1} f(x)\,dx$, $\int_{1}^{2} f(x)\,dx$, $\int_{0}^{2} f(x)\,dx$.

7.6   INTEGRATION BY DIRECT SUBSTITUTION

We now introduce various algebraic methods of transforming integrals to the standard forms of Table 7.5 so that they can be evaluated.

The process of integration by substitution involves making a change of variable so that the given integral is transformed into a standard form. The technique is very powerful.

Example 7J

Evaluate $\int (3x - 1)^4\,dx$.

Solution

The substitution $u = 3x - 1$ will change $(3x-1)^4$ to u^4. However $\int u^4\,dx$ involves two variables and so we must write dx in terms of du. We do this using the chain rule.

Now $u = 3x - 1$, so $\dfrac{du}{dx} = 3$ and hence $dx = \frac{1}{3}du$. So

$$\int (3x-1)^4 \, dx = \int u^4 \left(\tfrac{1}{3}du\right) = \tfrac{1}{3}\int u^4 \, du = \tfrac{1}{3}\left(\dfrac{u^5}{5}\right) + c$$

$$= \dfrac{1}{15}(3x-1)^5 + c.$$

Note that we write the answer in terms of x, not the changed variable u.

Example 7K

Evaluate $\int x(3x^2 + 2)^{-\frac{1}{2}} \, dx$.

Solution

The most difficult part of the function to be integrated is $(3x^2+2)^{-\frac{1}{2}}$. So we try $u = 3x^2 + 2$ as a substitution. So $\dfrac{du}{dx} = 6x$ and $x \, dx = \frac{1}{6}du$. Hence

$$\int x(3x^2 + 2)^{-\frac{1}{2}} \, dx = \int u^{-\frac{1}{2}}\left(\tfrac{1}{6}du\right) = \tfrac{1}{6}\int u^{-\frac{1}{2}} \, du = \tfrac{1}{6}(2u^{\frac{1}{2}}) + c$$

$$= \dfrac{1}{3}(3x^2 + 2)^{\frac{1}{2}} + c$$

Example 7L

Evaluate $\displaystyle\int \dfrac{4x+1}{2x^2 + x + 3}\,dx$.

Solution

Again the function to be integrated suggests the substitution $u = 2x^2 + x + 3$. So $\dfrac{du}{dx} = 4x+1$ and $(4x+1)dx = du$. Hence

$$\int \dfrac{4x+1}{2x^2 + x + 3}\,dx = \int \dfrac{1}{u}\,du = \ln u + c$$

$$= \ln(2x^2 + x + 3) + c.$$

▨▨▨▨▨▨▨▨▨▨▨ **Example 7M** ▨▨▨▨▨

Evaluate $\int_0^{\pi/2} \sin(2t + \pi)dt$.

Solution

For definite integrals we change variables and the limits to obtain a new definite integral.

Choose the substitution $u = 2t + \pi$ so that $\dfrac{du}{dt} = 2$ and $dt = \frac{1}{2}du$.

Now compute the new limits. For the lower limit $t = 0$, $u = \pi$ and for the upper limit $t = \dfrac{\pi}{2}$, $u = 2\pi$.

Hence changing variables

$$\int_0^{\pi/2} \sin(2t + \pi)dt = \int_\pi^{2\pi} \sin u(\tfrac{1}{2} du) = \frac{1}{2} \int_\pi^{2\pi} \sin u\, du$$

$$= \frac{1}{2}[-\cos u]_\pi^{2\pi}$$

$$= \frac{1}{2}(-\cos 2\pi + \cos \pi) = -1.$$

▨▨▨▨▨▨▨▨▨▨▨▨▨▨▨▨▨▨▨▨▨▨▨▨▨▨▨▨▨▨▨▨▨▨▨▨

All of these integrals can be evaluated using the TI-92.
Figure 7.22 shows the TI-92 home screen for these examples.

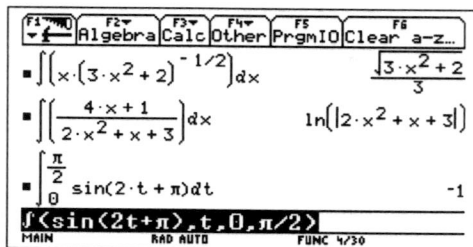

Figure 7.22

Exercise 7F

In each problem evaluate the integrals 'by hand' and then check your answers using the TI-92.

1.   Evaluate the following indefinite integrals.

(a) $\int x(4x^2 - 1)dx$

(b) $\int \sqrt{x - 1}dx$

(c) $\int \sin(2x + 1)dx$

(d) $\int \frac{2}{3x + 2}dx$

(e) $\int \frac{x}{5x^2 - 2}dx$

(f) $\int e^{2x+1} dx$

(g) $\int (4x + 1)^3 dx$

(h) $\int \frac{1}{(1+x)^2}dx$

(i) $\int \frac{1}{5x - 3}dx$

(j) $\int x e^{x^2} dx$

(k) $\int (1+x)^{0.3} dx$

(l) $\int \tan x dx$ (let $u = \cos x$)

(m) $\int (2x + 1)(x^2 + x + 3)^{0.5} dx$

(n) $\int \frac{2x + 3}{x^2 + 3x - 5}dx$

(o) $\int (4t - 11)^5 dt$

(p) $\int \frac{1}{(3 - 2v)^2}dv$

(q) $\int \cos(2y + \pi)dy$

(r) $\int \frac{t}{1 + t^2}dt$

(s) $\int \cos^2 \theta \sin\theta d\theta$

(t) $\int \sin^4(2\theta)\cos(2\theta)d\theta$

2.   Evaluate the following definite integrals.

(a) $\int_0^{\pi/2} \sin(3x - \pi)dx$

(b) $\int_1^3 \frac{1}{5x - 3}dx$

(c) $\int_1^2 \frac{1}{2t + 1}dt$

(d) $\int_1^3 \sqrt{2x - 1} \, dx$

(e) $\int_0^2 \frac{t^2}{1 + t^3}dt$

(f) $\int_0^1 x e^{x^2} dx$

(g) $\int_{-1}^2 e^{3u - 5} du$

(h) $\int_{-1}^1 (4v + 1)^3 dv$

7.7   INTEGRATION BY INDIRECT SUBSTITUTIONS

The substitutions in Section 7.7 are called **direct substitutions** because they tend to be fairly obvious from the function to be evaluated. Sometimes it is necessary to make trigonometric substitutions which may seem less than obvious. These are called **indirect substitutions**.

Example 7N

Evaluate $\int \dfrac{1}{\sqrt{16-x^2}} dx$.

Solution

Let $x = 4\sin\theta$ then $\dfrac{dx}{d\theta} = 4\cos\theta$ so $dx = 4\cos\theta d\theta$. Hence

$$\int \frac{1}{\sqrt{16-x^2}} dx = \int \frac{1}{\sqrt{16-16\sin^2\theta}} (4\cos\theta d\theta)$$

$$= \int \frac{1}{4\sqrt{1-\sin^2\theta}} (4\cos\theta d\theta) = \int \frac{1}{(4\cos\theta)} (4\cos\theta d\theta)$$

since $1 - \sin^2\theta = \cos^2\theta$

$$= \int 1 d\theta = \theta + c$$

$$= \arcsin\left(\frac{x}{4}\right) + c$$

Example 7P

Evaluate $\int \dfrac{1}{1+x^2} dx$.

Solution

Let $x = \tan\theta$ then $\dfrac{dx}{d\theta} = \sec^2\theta$ so $dx = \sec^2\theta d\theta$. Hence

$$\int \frac{1}{1+x^2}dx = \int \frac{1}{(1+\tan^2\theta)}(\sec^2\theta d\theta)$$

$$= \int \frac{1}{(\sec^2\theta)}(\sec^2\theta d\theta) \qquad \text{since } 1+\tan^2\theta = \sec^2\theta$$

$$= \int 1 d\theta = \theta + c$$

$$= \arctan x + c$$

As a general rule if the function to be integrated involves $a + bx^2$ then try $x = \sqrt{\frac{a}{b}}\tan\theta$; and if the function to be integrated involves $\sqrt{a-bx^2}$ then try $x = \sqrt{\frac{a}{b}}\sin\theta$. In the following exercise the substitutions are given.

Exercise 7G

Evaluate the following integrals using the suggested substitution. Check your answers using the TI-92.

(a) $\int \frac{1}{\sqrt{1-4x^2}}dx \qquad x = \frac{1}{2}\sin\theta$

(b) $\int \frac{1}{1+4t^2}dt \qquad t = \frac{1}{2}\tan\theta$

(c) $\int \frac{1}{\sqrt{9-x^2}}dx \qquad x = 3\sin\theta$

(d) $\int \frac{1}{1+9x^2}dx \qquad x = \frac{1}{3}\tan\theta$

(e) $\int_0^{\sqrt{3}/2} \frac{1}{1+4u^2}du \qquad u = \frac{1}{2}\tan\theta$

(f) $\int_1^{\sqrt{2}} \frac{1}{\sqrt{2-t^2}}dt \qquad t = \sqrt{2}\sin\theta$

7.8   A PAIR OF TRIGONOMETRIC INTEGRALS

There are two functions $\sin^2 x$ and $\cos^2 x$ whose integrals involve techniques which do not fit into the classification of substitutions.

These integrals are $\int \sin^2 x dx$ and $\int \cos^2 x dx$.

For these functions we use the double angle formulas

$$\cos 2x = 1 - 2\sin^2 x \quad \text{and} \quad \cos 2x = 2\cos^2 x - 1$$

$$\int \sin^2 x\,dx = \int \tfrac{1}{2}(1 - \cos 2x)dx = \frac{1}{2}x - \frac{1}{4}\sin 2x + c$$

$$\int \cos^2 x\,dx = \int \tfrac{1}{2}(1 + \cos 2x)dx = \frac{1}{2}x + \frac{1}{4}\sin 2x + c$$

7.9   INTEGRATION BY PARTS

In this section we introduce a rule for integrating products of functions. As with all rules in integration the method is not always guaranteed to be applicable in every case! We begin with the rule for differentiating a product

$$\frac{d}{dx}(uv) = u\frac{dv}{dx} + v\frac{du}{dx}$$

Integrating both sides with respect to x

$$\int \left[\frac{d}{dx}(uv)\right]dx = \int u\frac{dv}{dx}dx + \int v\frac{du}{dx}dx$$

so

$$uv = \int u\frac{dv}{dx}dx + \int v\frac{du}{dx}dx$$

Rearranging gives

$$\boxed{\int u\frac{dv}{dx}dx = uv - \int v\frac{du}{dx}dx}$$

This rule can be used to integrate some products of functions. It changes the integral on the left hand side to a product of two functions and another integral on the right hand side. The method is called **integration by parts** and the aim is to make the new integral easier than the original one. The following example shows how the method works.

Example 7Q

Evaluate $\int x\cos x\,dx$.

Solution

We have to integrate the product of two functions x and $\cos x$.
Suppose that we let $u = x$ and $\dfrac{dv}{dx} = \cos x$. Then $v = \sin x$ and $\dfrac{du}{dx} = 1$. So

$$\int u \frac{dv}{dx}\,dx = uv - \int v \frac{du}{dx}\,dx \quad \text{becomes}$$

$$\int x \cos x\,dx = x \sin x - \int (\sin x)(1)dx$$

The new integral is quite straightforward. Hence

$$\int x \cos x\,dx = x \sin x + \cos x + c$$

In evaluating the integral we made the choice $u = x$. Suppose we had let $u = \cos x$ and $\dfrac{dv}{dx} = x$. Then $v = \frac{1}{2}x^2$ and $\dfrac{du}{dx} = -\sin x$. So

$$\int x \cos x\,dx = \frac{1}{2}x^2 \cos x - \int \frac{1}{2}x^2(-\sin x)dx$$

$$= \frac{1}{2}x^2 \cos x + \frac{1}{2}\int x^2 \sin x\,dx .$$

Clearly the new integral is harder than the original one. So the substitution for u and $\dfrac{dv}{dx}$ needs to be made carefully.

Example 7R

Evaluate $\displaystyle\int x^2 e^{-3x}\,dx$.

Solution

The choice here is $u = x^2$ because after differentiating x^2 twice this will give a constant.
Let $u = x^2$ and $\dfrac{dv}{dx} = e^{-3x}$. Then $\dfrac{du}{dx} = 2x$ and $v = -\frac{1}{3}e^{-3x}$.

$$\int u\frac{dv}{dx}\,dx = uv - \int v\frac{du}{dx}\,dx \quad \text{becomes}$$

$$\int x^2\,\mathrm{e}^{-3x}\,dx = -\frac{1}{3}x^2\,\mathrm{e}^{-3x} - \int\left(-\frac{1}{3}\mathrm{e}^{-3x}\right)(2x)\,dx$$

$$= -\frac{1}{3}x^2\,\mathrm{e}^{-3x} + \frac{2}{3}\int x\mathrm{e}^{-3x}\,dx . \qquad *$$

The new integral must be evaluated in parts. Let $u = x$ and $\dfrac{dv}{dx} = \mathrm{e}^{-3x}$. Then $\dfrac{du}{dx} = 1$ and $v = -\frac{1}{3}\mathrm{e}^{-3x}$. So

$$\int x\mathrm{e}^{-3x}\,dx = -\tfrac{1}{3}x\mathrm{e}^{-3x} - \int\left(-\tfrac{1}{3}\mathrm{e}^{-3x}\right)(1)\,dx$$

$$= -\tfrac{1}{3}x\mathrm{e}^{-3x} + \tfrac{1}{3}\int\mathrm{e}^{-3x}\,dx$$

$$= -\tfrac{1}{3}x\mathrm{e}^{-3x} - \tfrac{1}{9}\mathrm{e}^{-3x} + c$$

Substituting into * gives

$$\int x^2\,\mathrm{e}^{-3x}\,dx = -\frac{1}{3}x^2\,\mathrm{e}^{-3x} + \frac{2}{3}\left[-\frac{1}{3}x\mathrm{e}^{-3x} - \frac{1}{9}\mathrm{e}^{-3x} + c\right]$$

$$= -\left(\frac{1}{3}x^2 + \frac{2}{9}x + \frac{2}{27}\right)\mathrm{e}^{-3x} + A$$

where A is an arbitrary constant.

████████████████ **Exercise 7H** ████████████████████████████████████

In each problem evaluate the integrals 'by hand' and then check your answers using the TI-92.

1.   Evaluate the following indefinite integrals.

(a) $\int x \sin x dx$ (b) $\int x e^x dx$

(c) $\int x^2 e^{2x} dx$ (d) $\int x \sin 3x dx$

(e) $\int t^2 e^t dt$ (f) $\int x \ln x dx$

(g) $\int \ln x dx$ [Hint: let $u = \ln x$ and $\dfrac{dv}{dx} = 1$].

2.   Evaluate the following definite integrals.

(a) $\int_0^1 u e^u du$ (b) $\int_0^{\pi/2} x \cos 2x dx$

(c) $\int_2^4 x\sqrt{x-1} dx$ (d) $\int_0^2 t^2 e^{-2t}$

(e) $\int_0^{\pi/2} e^x \sin x dx$ (f) $\int_0^{\pi/3} e^{-2x} \sin 3x dx$

7.10   USE OF PARTIAL FRACTIONS

The integral of the form $\int \dfrac{1}{a+bx} dx$ can be evaluated to give $\dfrac{1}{b}\ln(a+bx)$ and this provides a standard form for evaluating integrals of functions given as $f(x)/g(x)$ where $f(x)$ and $g(x)$ are polynomials. The technique is to transform the integral using a partial fraction expansion. The following example illustrates the approach.

████████████████ **Example 7S** ████████████████████████████████████

Evaluate $\int \dfrac{x+1}{x^2+x-2} dx$.

Solution

The factors of $x^2 + x - 2$ are $(x+2)$ and $(x-1)$, so we write the function in the form

$$\frac{x+1}{x^2+x-2} = \frac{x+1}{(x+2)(x-1)} = \frac{A}{(x+2)} + \frac{B}{(x-1)}$$

This is called the **partial fraction** form.
    If we combine the two **partial fractions** on the right hand side we get

$$\frac{A}{(x+2)} + \frac{B}{(x-1)} = \frac{A(x-1) + B(x+2)}{(x+2)(x-1)}$$

Comparing the numerators of the first and last fractions leads to

$$x + 1 = A(x-1) + B(x+2)$$

To find the values of A and B first, let $x = 1$ to give $2 = 0 + 3B$, so that $B = \frac{2}{3}$. Then let $x = -2$ to give $-1 = -3A + 0$, so that $A = \frac{1}{3}$.
    Alternatively we could compare coefficients of x on each side of the equation. Since

$$x + 1 = (A+B)x + 2B - A$$

then

$$A + B = 1 \quad \text{and} \quad 2B - A = 1$$

Solving these gives $A = \frac{1}{3}$ and $B = \frac{2}{3}$ as before. Now the integral can be written as

$$\int \frac{x+1}{x^2+x-2}dx = \int \frac{\frac{1}{3}}{(x+2)} + \frac{\frac{2}{3}}{(x-1)}dx$$

$$= \frac{1}{3}\ln(x+2) + \frac{2}{3}\ln(x-1) + c$$

This method of approach can be applied provided the denominator $g(x)$ has real linear factors. However there are many polynomials with quadratic factors $ax^2 + bx + c$ which cannot be factorized into two linear factors. This then requires evaluation of integrals of the form

$$\int \frac{1}{ax^2 + bx + c} dx$$

The TI-92 gives partial fractions directly using 3:expand(on the **F2** menu. Check the partial fraction expansion in Example 7R using the TI-92.

Example 7T

Evaluate $\int \frac{1}{x^2 - 4x + 6} dx$.

Solution

The first step is to consider the function $x^2 - 4x + 6$ and complete the square

$$x^2 - 4x + 6 = (x - 2)^2 + 2$$

Now let $u = x - 2$ so that $dx = du$. On substitution the integral becomes

$$\int \frac{1}{x^2 - 4x + 6} dx = \int \frac{1}{(x - 2)^2 + 2} dx$$

$$= \int \frac{1}{u^2 + 2} du .$$

Now with $u = \sqrt{2} \tan\theta$ and $du = \sqrt{2} \sec^2 \theta d\theta$ this integral becomes

$$\int \frac{1}{u^2 + 2} du = \int \frac{1}{2(1 + \tan^2 \theta)} \sqrt{2} \sec^2 \theta d\theta$$

$$= \int \frac{1}{\sqrt{2}} d\theta = \frac{\theta}{\sqrt{2}} + c$$

Replacing θ by $\arctan(u / \sqrt{2})$ and u by $x - 2$ gives

$$\int \frac{1}{x^2 - 4x + 6} dx = \frac{1}{\sqrt{2}} \arctan\left((x - 2) / \sqrt{2}\right) + c$$

Exercise 7I

Evaluate the following integrals. Check your results using the TI-92.

(a) $\displaystyle\int \frac{2}{(x+1)(x-1)}\,dx$

(b) $\displaystyle\int \frac{1}{(2x-1)(x+2)}\,dx$

(c) $\displaystyle\int \frac{5}{(x+1)(x-3)}\,dx$

(d) $\displaystyle\int \frac{3}{(t-2)(t-3)}\,dt$

(e) $\displaystyle\int \frac{2v+1}{(2v-1)(2v+3)}\,dv$

(f) $\displaystyle\int_1^2 \frac{2x+1}{(x+2)(x+3)}\,dx$

(g) $\displaystyle\int \frac{2x^2-9x+1}{x(x-1)(x+3)}\,dx$

(h) $\displaystyle\int_0^1 \frac{t(t-2)}{(t+1)(t^2+1)}\,dt$

(i) $\displaystyle\int \frac{1}{x^2-9x+25}\,dx$

(j) $\displaystyle\int_0^1 \frac{1}{(x+1)^2(x+2)}\,dx$

Exercise 7J

Integration is a vast topic which involves several different techniques. The skill is deciding which technique to use. The following integrals will give you practice at choosing the right method. Check your answers using the TI-92.

1.   Evaluate the following integrals.

(a) $\displaystyle\int \sin x \cos x\,dx$

(b) $\displaystyle\int \frac{3x^2}{x^3-3}\,dx$

(c) $\displaystyle\int e^{2x} - \sin 4x\,dx$

(d) $\displaystyle\int_1^4 \frac{dx}{\sqrt{5-x}}$

(e) $\displaystyle\int x \sin^2 x\,dx$

(f) $\displaystyle\int_0^1 \frac{1}{4+t^2}\,dt$

(g) $\displaystyle\int_0^1 \frac{1}{4-t^2}\,dt$

(h) $\displaystyle\int \frac{t}{4+t^2}\,dt$

(i) $\displaystyle\int_0^1 t\,e^{4t^2}\,dt$

(j) $\displaystyle\int \frac{5}{t^2+t-6}\,dt$

(k) $\displaystyle\int \frac{u^2+2}{u}\,du$

(l) $\displaystyle\int \sqrt{1-v^2}\,dv$

(m) $\int u^2(u^2+1)du$ (n) $\int_0^2 e^{(4t-1)}dt$ (o) $\int u^2 \sin u\, du$

(p) $\int t e^{3t}\, dt$

2. The speed of an object travelling in a straight line is given by $(3t+5)^2$ ms^{-1} where t is the time. Find the formula for the distance travelled in the first 5 seconds.

3. In a physics experiment it is found that the rate of change of temperature is inversely proportional to $(2t+3)^2$. Find a formula for the temperature as a function of time.

4. The force acting on an object due to an elastic spring is given by $F(x) = 15x - 20$. The potential energy $V(x)$ is related to the force by $\dfrac{dV}{dx} = F$. Use integration to find a formula for $V(x)$.

5. The gradient of a curve at any point is given by $\dfrac{dy}{dx} = \dfrac{1}{x^2 - 3x + 2}$. If the curve passes through the point $(0,1)$ find the equation of the curve.

6. Find the total area enclosed by the curve $y = x(x-1)(x-2)$ and the x-axis between $x = 0$ and $x = 4$.

7. Find the area enclosed by the line $y = 2$ and the curve $y = x(3-x)$.

8. A function $f(t)$ is given in tabular form as

t	$f(t)$
1.8	6.050
2.0	7.389
2.2	9.025
2.4	11.023
2.6	13.464
2.8	16.445
3.0	20.086
3.2	24.533
3.4	29.964

Find an approximate value for $\int_{1.8}^{3.4} f(t)dt$.

9.   Consider the function $f(x) = 2x - 3$.

     (a)   Draw a graph of the function and find the area between $f(x)$, the x-axis, $x = 1$ and $x = 2$.

     (b)   Evaluate $\int_1^2 (2x - 3)dx$. Are your answers to (a) and (b) equal? Does an integral always equal an area?

10.   An oil droplet is falling through a medium which produces a resistance proportional to $v^{3/2}$ where v is its downward velocity. The time taken for it to reach a velocity of $u/2$, where u is its terminal velocity, is given by

$$t = \frac{u}{g} \int_0^{0.5} \frac{1}{1 - x^{3/2}} dx$$

Evaluate this expression using the TI-92.

11.   The fraction of internally radiated heat energy falling on a thermocouple which is centrally placed on the axis of a hot tube of length L and radius R is given by

$$H = \frac{1}{2} \int_{-1}^{1} \frac{a}{\left[1 + (ax)^2\right]^{3/2}} dx$$

where $a = \dfrac{L}{2R}$. Find the value of H for a tube of length 8 cm and radius 8 cm.

8

Numerical methods

8.1 NEWTON-RAPHSON METHOD

In Chapter 5 you will have seen how to use iterative formulae to find the solutions of equations. In this section a method for formulating iterative formulae that converge will be developed to solve equations of the form $f(x) = 0$.

TI-92 ACTIVITY 8A

Clear the Y = Editor and GRAPH screens.

(A) Consider the equation

$$x^3 + 2x - 2 = 0$$

(i) Set $y1 = x^3 + 2x - 2$ and GRAPH $y1$ using the WINDOW

xmin = –0.5, xmax = 2, xscl = 0.1,
ymin = –1, ymax = 5, yscl = 0.5.

(ii) This investigation shows how to improve the accuracy of solutions by using the TI-92 Tangent command. Take $x = 1.5$ as a first estimate of the solution of $x^3 + 2x - 2 = 0$.
Draw the tangent at $x = 1$ (**F5** select A:Tangent). Note that the tangent crosses the x-axis at a point closer to the curve than the original estimate of $x = 1.5$. The equation of the tangent is $y = 8.75x - 8.75$. It cuts the x-axis when $y = 0$. Solving $8.75x - 8.75 = 0$ gives $x = 1$ as an improved estimate.

(iii) Find the equation of the tangent to the curve at the new estimate of the solution, in this case $x = 1.0$. From the equation of the tangent find the improved estimate of the solution of the equation, you should get 0.8. Repeat step (iii) and find an improved estimate.

Figure 8.1 shows the screen dump after drawing three tangents.

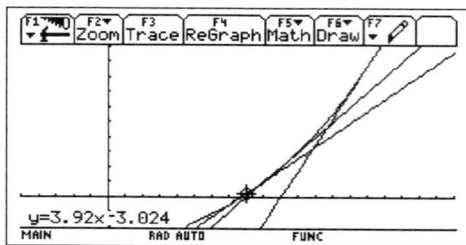

Figure 8.1   Using the tangent facility to solve $x^3 + 2x - 2 = 0$

The estimate to the solution after three tangents is $\dfrac{3.024}{3.92} = 0.7714$.

(B)   Repeat the activity for the equation

$$x^4 - 1.4x^3 - 1.8x^2 + x = 0$$

(C)   By drawing a tangent to the curve at the first estimate given below, find an improved estimate of the solution.

(i) $\dfrac{x^4}{10} - 1 = 0$, first estimate $x = 2$.

(ii) $\cos x - x = 0$, first estimate $x = 1$.

(iii) $e^x - x^2 = 0$, first estimate $x = -1$.

From the TI-92 Activity 8A you will have seen how by repeatedly drawing tangents it is possible to move closer and closer to the solution. You may also have noticed that only a few steps are needed to produce a good result compared with the methods used in Chapter 5.

Figure 8.2 shows the graph of a function $f(x)$, and how the tangent to the curve gives an improved estimate x_{n+1} from a previous estimate x_n.

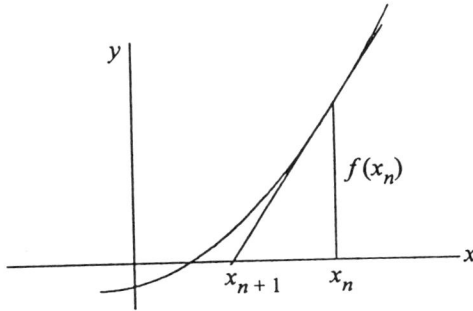

Figure 8.2

The gradient of the tangent at $x = x_n$ is given by $f'(x_n)$. It is also possible to find the gradient of the tangent by examining the triangle shown in Figure 8.2. The height of the curve at x_n is $f(x_n)$, so the slope of the tangent is given by

$$\frac{f(x_n)}{x_n - x_{n+1}}$$

These two expressions can now be equated to give

$$f'(x_n) = \frac{f(x_n)}{x_n - x_{n+1}}$$

This can be rearranged to give

$$x_n - x_{n+1} = \frac{f(x_n)}{f'(x_n)}$$

or

$$x_{n+1} = x_n - \frac{f(x_n)}{f'(x_n)}$$

This iterative scheme is known as the **Newton-Raphson method** and provides a powerful method for finding the solutions of equations.

▨▨▨▨▨▨▨▨▨▨ **Example 8A** ▨▨▨▨▨▨▨▨▨▨▨▨▨▨▨▨▨▨▨▨▨▨▨▨

Find the positive solution of $x^4 - x^2 - 1 = 0$.

Solution

Before the Newton-Raphson method can be applied a first estimate of the solution needs to be found. A sign-search procedure can be used to do this. Table 8.1 below shows this.

Table 8.1

x	0	1	2
$x^4 - x^2 - 1$	-1	-1	11

A solution of $x^4 - x^2 - 1 = 0$ lies between $x = 1$ and $x = 2$. As the sign changes between $x = 1$ and $x = 2$, it is sensible to choose a value in this range as the first estimate. In this case we choose $x_1 = 1$.

Here $f(x)$ is defined as

$$f(x) = x^4 - x^2 - 1$$

and differentiating gives

$$f'(x) = 4x^3 - 2x$$

So the Newton-Raphson method

$$x_{n+1} = x_n - \frac{f(x_n)}{f'(x_n)}$$

gives the iterative formula

$$x_{n+1} = x_n - \frac{(x_n^4 - x_n^2 - 1)}{(4x_n^3 - 2x_n)}$$

Taking $x_1 = 1$ we can calculate $x_2, x_3,....$

$$x_2 = 1 - \frac{(1^4 - 1^2 - 1)}{(4 \times 1^3 - 2 \times 1)} = 1.5$$

$$x_3 = 1.5 - \frac{(1.5^4 - 1.5^2 - 1)}{(4 \times 1.5^3 - 2 \times 1.5)} = 1.327$$

Repeating the process leads to the values

$$x_4 = 1.276$$
$$x_5 = 1.272$$
$$x_6 = 1.272$$

The positive solution of the equation $x^4 - x^2 - 1 = 0$ is 1.272 to 3 decimal places.

TI-92 ACTIVITY 8B

Clear the HOME screen. Select APPROX mode.

(A)   Use the TI-92 to solve the equation $x^2 = e^x$ by following the procedure below.

(i)   Rewrite $x^2 = e^x$, in the form $x^2 - e^x = 0$, so that $f(x) = x^2 - e^x$.

(ii)   Graph this function. Read off a first solution of the equation.

(iii)   Differentiate this function.

(iv)   The Newton-Raphson formula for solving $x^2 - e^x$ has the iterative function

$$x - \frac{(x^2 - e^x)}{(2x - e^x)}$$

Find an improved estimate using the first estimate taken from the graph in (ii) and the Newton-Raphson formula.

(v)   Repeat (iv) until you can give the solution correct to 6 decimal places.

(vi)   Now we introduce a method of doing the iterations automatically.

Input −1 **ENTER**

type ans(1) − ((ans(1))^2 − e^(ans(1)))/(2*ans(1)

− e^(ans(1))) **ENTER**

This is the formula

$$\text{ans}(1) \ - \ \frac{(\text{ans}(1))^2 - e^{\text{ans}(1)})}{2\text{ans}(1) - e^{\text{ans}(1)}}$$

ans(1) takes the place of x in order to achieve an iterative process.
In typing −1 first, this starts the sequence at −1 i.e. ans(1) = −1.
The TI-92 treats ans(1) as whatever is in the last line that was input on
the HOME screen.
Figure 8.3 shows the HOME screen for this activity.

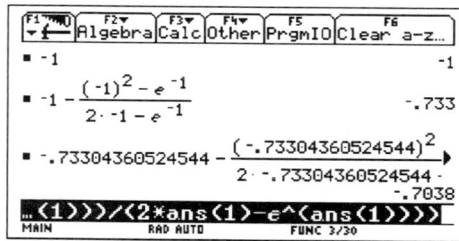

Figure 8.3   The Newton-Raphson iteration

Continue to press **ENTER** until you have the answer to 6 decimal
places. Notice how quickly the method converges to a solution
−0.703467. Verify that this solution satisfies the original equation.

(vii)   Try using other starting points. Do they all converge to the same value?

(B)   Consider the equation $x^3 - 5x^2 + 4x + 2 = 0$.

(i)   Set up the Newton-Raphson formula for this equation in the TI-92.

(ii)   Find solutions by using the method in (A) (vi), for each of the following
first estimates.

(a)   0   (b)   1   (c)   5   (d)   2.7

(iii)   Graph $x^3 - 5x^2 + 4x + 2$ and check that the 3 solutions you obtained in (ii) are reasonable.

(iv)   It seems strange that by starting at 2.7 you obtain the negative solution. To see why this happens use the Tangent command to find the equation of the tangent to the curve at $x = 2.7$. Plot this tangent and explain why it leads to the negative solution.

(C)   Set up the Newton-Raphson formula in the TI-92 to solve the following equation.

$$x^3 - 3x - 5 = 0$$

Graph the appropriate function $f(x)$. Calculate the values of x for which $f'(x) = 0$. Draw the tangent to the graph at these values of x. Explain why these values should be avoided as first estimates in the Newton-Raphson formula.

(D)   Repeat (C) for

(i) $x^2 - 2\sin x = 0$

(ii) $x - e^{x^2} = 0$

In each case give ranges of values of x for first estimates to use in the Newton-Raphson formula.

Summary

The Newton Raphson iterative formula converges to a limit very quickly for most initial values. But there can be problems! Particular care is needed in the choice of the first guess x_1. Consider the function in Figure 8.4.

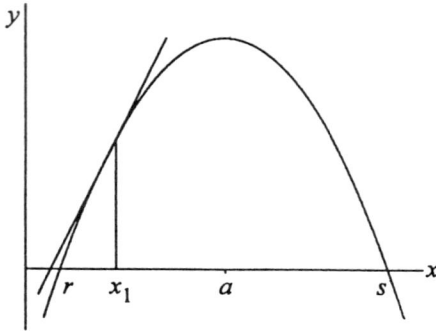

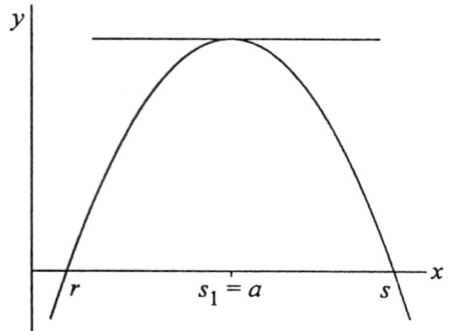

Figure 8.4   Figure 8.5

Suppose we wish to find the solution at $x = r$ and there is a local maximum at $x = a$. If we choose our starting iterate x_1 between r and a then the Newton-Raphson method will converge to $x = r$. However if we choose x_1 between a and s the method will converge to $x = s$ and not the root that we require.

Clearly if we choose $x_1 = a$, the turning point, then we are in trouble because the tangent is parallel to the x-axis (see Figure 8.5). Finally if $\frac{df}{dx} = 0$ at the root then the Newton-Raphson method is not an appropriate method to use.

The TI-92 has a built in function for solving non-linear equations, it is called nSolve. This facility iteratively searches for one approximate real solution of an equation. For example, to solve $x - e^{-x} = 0$ we type

$$\text{nSolve}(x - e\char94(-x) = 0, x) \text{ ENTER}$$

The solution is 0.567143 to six decimal places. Try this function to check your answers in Exercise 8A.

Exercise 8A

1.   Use the Newton-Raphson method to solve the equations

(a) $x - 2\sin x = 0$ (b) $x^3 - x + 8 = 0$

(c) $x^3 = e^x$ (d) $\cos x = x$

(e) $x^4 = x + 2$ (f) $\frac{1}{x} = e^x$

2. By writing $x^n = a$, find an iterative formula to find the nth root of a number. Use it to find

    (a) $\sqrt[3]{10}$ (b) $\sqrt[5]{8}$ (c) $\sqrt[10]{100}$

3. Consider the equation $x^3 - 3x = 0$.

    (a) Graph the curve $y = x^3 - 3x$.

    (b) There are two values of x for which the Newton-Raphson formula does not converge. Identify these points by drawing tangents and then explain why the method fails.

4. Find the positions of the turning points of the curve $y = e^x - 4x^2$.

5. A motor under load generates heat at a constant rate and radiates it at a rate proportional to the excess temperature T, so that at time t (in minutes)

    $$T = \frac{10}{k} - \frac{10}{k}e^{-kt}$$

    where k is a positive constant. After 10 minutes the temperature rise is 50°C. Find the value of k to four decimal places using the Newton-Raphson method.

8.2 APPROXIMATING FUNCTIONS BY A SERIES

TI-92 ACTIVITY 8C

(A) (i) Graph the function $\sin x$. Then graph the series

    $$x - \frac{x^3}{6} + \frac{x^5}{120} - \frac{x^7}{5040}$$

     (ii) Compare the two graphs. What properties do they have in common? For what range of values does the series provide a good approximation to $\sin x$?

(B) (i) Graph the function e^x. Then graph the series

    $$1 + x + \frac{x^2}{2} + \frac{x^3}{6} + \frac{x^4}{24}$$

(ii)   Again compare the two graphs. What features do they have in common? For what range of values does the series provide a good approximation to e^x?

(C)   Each of the series below approximate one of the functions A, B and C.

A $\dfrac{1}{x}$ B $\cos(x)$ C $\sin(2x)$

(i) $1 - \dfrac{x^2}{2} + \dfrac{x^4}{24} - \dfrac{x^6}{720}$ (ii) $6 - 15x + 20x^2 - 15x^3 + 6x^4 - x^5$

(iii) $2x - \dfrac{4x^3}{3} + \dfrac{4x^5}{15} - \dfrac{8x^7}{315}$

Graph each series. Then try to identify the function. Graph the function to check your prediction.

Over what range of values is each approximation reasonable?

8.3   THE MACLAURIN SERIES

Some of the series considered above such as e^x, $\sin x$ and $\cos x$ gave their best approximations when x was close to zero, but as x increased or decreased, the goodness of the approximation would decrease. Such approximations are known as **Maclaurin series** and are based on information about the function at $x = 0$.

To form such a series consider an approximation of the form of a polynomial

$$f(x) = a_0 + a_1 x + a_2 x^2 + a_3 x^3 + a_4 x^4 + \cdots$$

where a_i are coefficients that have to be determined to give a good approximation.

When $x = 0$, then $f(0) = a_0$ as all the other terms will be zero.

Differentiating the series four times gives

$$f'(x) = a_1 + 2a_2 x + 3a_3 x^2 + 4a_4 x^3 + \cdots$$
$$f''(x) = 2a_2 + 6a_3 x + 12a_4 x^2 + \cdots$$
$$f'''(x) = 6a_3 + 24a_4 x + \cdots\cdots$$
$$f^{iv}(x) = 24a_4 + \cdots\cdots\cdots$$

Now substituting $x = 0$ into each of these expressions gives

$$f'(0) = a_1$$
$$f''(0) = 2a_2$$
$$f'''(0) = 6a_3$$
$$f^{iv}(0) = 24a_4$$

So the coefficients a_i are determined by the values of $f(x)$ and its derivatives when $x = 0$.

$$a_0 = f(0) \quad a_1 = f'(0) \quad a_2 = \frac{f''(0)}{2} \quad a_3 = \frac{f'''(0)}{6} \quad a_4 = \frac{f^{iv}(0)}{24}$$

In general

$$a_n = \frac{f^{(n)}(0)}{n!}$$

So the Maclaurin series can be written as

$$f(x) = f(0) + f'(0)x + \frac{f''(0)x^2}{2!} + \frac{f'''(0)x^3}{3!} + \cdots + \frac{f^n(0)x^n}{n!} + \cdots$$

This is in fact a series with an infinite number of terms, but in practice only a few of the first terms are used to produce the approximations. For this reason the approximation is only valid for values of x close to 0.

Example 8B

Find the Maclaurin series for $\sin x$, giving the first 3 non-zero terms of the series.

Solution

The first step is to find the derivatives of $\sin x$ and evaluate each of these and $\sin x$ at $x = 0$, as shown below.

$$f(x) = \sin x \qquad f(0) = 0$$
$$f'(x) = \cos x \qquad f'(0) = 1$$
$$f''(x) = -\sin x \qquad f''(0) = 0$$
$$f'''(x) = -\cos x \qquad f'''(0) = -1$$
$$f^{iv}(x) = \sin x \qquad f^{iv}(0) = 0$$
$$f^{v}(x) = \cos x \qquad f^{v}(0) = 1$$

As there are now 3 non-zero derivatives it is possible to form the Maclaurin series. The series is defined as

$$f(x) = f(0) + f'(0)x + \frac{f''(0)x^2}{2!} + \frac{f'''(0)x^3}{3} + \frac{f^{iv}(0)x^4}{4!} + \frac{f^{v}(0)x^5}{5!} + \cdots$$

In this case we obtain

$$f(x) = 0 + 1 \times x + \frac{0 \times x^2}{2!} + \frac{(-1) \times x^3}{3!} + \frac{0 \times x^4}{4!} + \frac{1 \times x^5}{5!} + \cdots$$

i.e.

$$\sin x = x - \frac{x^3}{6} + \frac{x^5}{120} + \cdots$$

Exercise 8B

1. Find the Maclaurin series of the following functions giving the first three non-zero terms.

    (a) e^x (b) $\cos x$

    (c) $\sin(x^2)$ (d) $\dfrac{1}{1+x}$

    (e) $\tan x$ (f) $\sqrt{1-x}$

    (g) $\ln(\cos x)$ (h) e^{x^2}

    Use the TI-92 to compare each function with its Maclaurin series and in each case suggest the range of values of x that the approximation is a good one.

    For parts (d) and (f) compare your Maclaurin series with the series obtained using the binomial theorem.

2.   Use the Maclaurin series expansion of $(1+x)^n$ to obtain the binomial theorem

$$(1+x)^n = 1+nx+\frac{n(n-1)x^2}{2!}+\frac{n(n-1)(n-2)x^3}{3!}+\cdots$$

Use the TI-92 the graph to plot $(1+x)^n$ and its Maclaurin series for $n = 0.5$. How many terms do you need to take for the graphs to be the same between $x = 0$ and $x = 1$?

8.4   TAYLOR SERIES

The Maclaurin series for a function is valid for values of x close to zero and finding it depends on you being able to calculate the value of the function and its derivatives at zero. For a function such as $\ln(x)$ it is impossible to find a Maclaurin series because neither the function or its derivatives are defined when $x = 0$. A *Taylor series* is similar to the Maclaurin series, but is valid close to values of x other than zero. The Maclaurin series is actually a special case of the Taylor series.

The form of a Taylor series about a is

$$f(x) = a_0 + a_1(x-a) + a_2(x-a)^2 + a_3(x-a)^3 + \cdots$$

To obtain the values of the coefficients a_0, a_1, etc, we differentiate repeatedly, in the same way as for the Maclaurin series, and substitute in $x = a$ to the function and its derivatives; differentiating

$$f'(x) = a_1 + 2a_2(x-a) + 3a_3(x-a)^2 + \cdots$$
$$f''(x) = 2a_2 + 6a_3(x-a) + \cdots$$
$$f'''(x) = 6a_3 + \cdots$$

Substituting $x = a$ gives

$$a_0 = f(a)$$

$$a_1 = f'(a)$$

$$a_2 = \frac{f''(a)}{2!}$$

$$a_3 = \frac{f'''(a)}{3!}$$

$$a_n = \frac{f^n(a)}{n!}$$

So the Taylor series is given by

$$f(x) = f(a) + f'(a)(x-a) + \frac{f''(a)(x-a)^2}{2!} + \frac{f'''(a)(x-a)^3}{3!} + \cdots$$

Notice that if $a = 0$ this reduces to the Maclaurin series.

Example 8C

(i)   Find the first three non-zero terms of the Taylor series expansion for $\ln(x)$ about $x = 1$.

(ii)   Use the series to find $\ln(1.1)$.

Solution

(i)   The first step is to obtain the derivatives of the function and substitute $x = 1$ into these and the functions

$$f(x) = \ln(x) \qquad\qquad f(1) = \ln(1) = 0$$

$$f'(x) = \frac{1}{x} \qquad\qquad f'(1) = \frac{1}{1} = 1$$

$$f''(x) = -\frac{1}{x^2} \qquad\qquad f''(1) = -\frac{1}{1^2} = -1$$

$$f'''(x) = \frac{2}{x^3} \qquad\qquad f'''(1) = \frac{2}{1^3} = 2$$

The Taylor series about $x = 1$ is then given by

$$f(x) = f(1) + f'(1)(x-1) + \frac{f''(1)(x-1)^2}{2!} + \frac{f'''(1)(x-1)^3}{3!} + \cdots$$

$$= 0 + 1(x-1) + \frac{(-1)(x-1)^2}{2} + \frac{2(x-1)^3}{6} + \cdots$$

$$= (x-1) - \frac{(x-1)^2}{2} + \frac{(x-1)^3}{3} + \cdots$$

(ii)   To evaluate $\ln(1.1)$ simply substitute $x = 1.1$ into the Taylor series

$$\ln(1.1) = (1.1-1) - \frac{(1.1-1)^2}{2} + \frac{(1.1-1)^3}{3} + \cdots$$

$$= 0.1 - \frac{0.1^2}{2} + \frac{0.1^3}{3} + \cdots$$

$$= 0.1 - 0.005 + 0.0003$$

$$= 0.0953 \,.$$

TI-92 ACTIVITY 8D

Clear the HOME screen. Select EXACT mode. It is straightforward to obtain a Taylor or Maclaurin series with the TI-92. This process is first described and then used to illustrate some of the properties of these series.

(A)   (i)   Press **F3** and select 9:taylor(**ENTER**.

type $\sin(x),x,5,0)$ **ENTER**

The entry 5 denotes the highest power in the series and 0 is the point about which the series is expanded.
Explain why the series you obtain is a Maclaurin series.

(ii)   Now plot the original function and its Maclaurin series. What can you say about the range of validity of the approximation?

(iii)   Investigate how including more terms improves the range of validity for the approximation.

(iv)   What happens if you expand about $\pi/2$, π or 2π rather than 0?

(B)   (i)   Find the Maclaurin series for e^x. Plot it and compare it with the original function.

(ii)   Investigate how increasing the number of terms improves the quality of the approximation.

(iii)   Try using a Taylor series expanded with a negative value for a. How does this affect the goodness of the approximation?

(C)   Try to find a series approximation for $\ln(x)$. What problems do you encounter?

(D)   Find the Taylor series for $\dfrac{1}{1+x}$ expanded about $x = -0.5$ and $x = 0.5$.

Compare the graph of the function and the graph of the first four non-zero terms of each of the Taylor series. Suggest a range of values of x for which the series expansion is a good one. Are there any values of x for which a Taylor series expansion of $\dfrac{1}{1+x}$ does not exist?

Exercise 8C

1.   Find the first 3 non-zero terms of the Taylor series for

   (a) $\cos(x)$ expanded about $\dfrac{\pi}{3}$,

   (b) $\tan(x)$ expanded about $\dfrac{\pi}{4}$,

   (c) $\dfrac{1}{x}$ expanded about 1.

2.   (a)   Find the Maclaurin series for $\sin(2x)$.

   (b)   Write down the Maclaurin series for $\sin(3x)$.

3.   (a)   Explain why it is possible to find a Maclaurin series for $\ln(\cos x)$, but not for $\ln(\sin x)$.

   (b)   Find the first 5 non-zero terms of the series for $\ln(\cos x)$.

   (c)   Use your series to find an approximation for

$$\int_0^{0.1} \frac{\ln(\cos x)}{x^2}\,dx$$

4.   Use a Maclaurin series to help you find an approximation for

$$\int_{0.5}^{1} \frac{e^x}{x}\,dx$$

What criteria did you use for your choice of degree of the Maclaurin series expansion of e^x?

5.   Use a Maclaurin series to help you find an approximation for

$$\int_{0}^{1} e^{x^2}\,dx$$

What criteria did you use for your choice of degree of the Maclaurin series expansion of e^{x^2}?

8.5   NUMERICAL INTEGRATION

The direct and indirect (by substitution) methods of integration described in Chapter 7 provide exact, usually called **analytical**, solutions of integrals. Often in scientific and technological applications we need to evaluate the integral of a function which is given in tabular form so that the function to be integrated is not known. We need a method of integration just using these numerical values. Furthermore, it is often necessary to integrate a function for which the analytical methods do not work. For example, $\int e^{x^2}\,dx$ does not have an analytical solution.

Considering integrals as areas under graphs allows us to develop approximate methods of integration. In the introduction to integration the areas under graphs were estimated by means of summing the areas of rectangles (see section 7.1). An alternative approach is to use trapeziums.

Consider the integral $\int_{a}^{b} f(x)\,dx$ and suppose that we divide the interval [a,b] in n equal subintervals each of width $h = (b-a)/n$. Approximate the curve by a set of trapeziums as shown.

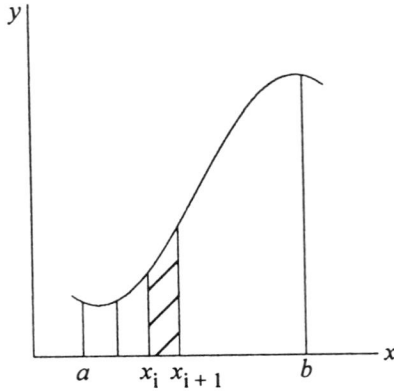

Figure 8.6

The area of the trapezium shown shaded is

$$\frac{1}{2}\left[f(x_i) + f(x_{i+1})\right](x_{i+1} - x_i)$$

The total area of all the trapeziums approximating the curve is

$$s = \sum_{i=0}^{n-1} \frac{1}{2}h\left[f(x_i) + f(x_{i+1})\right]$$

where $x_0 = a$ and $x_n = b$.

If we write out some of these terms we can simplify this formula.

$$s = \frac{1}{2}h\left\{\left[f(x_0) + f(x_1)\right] + \left[f(x_1) + f(x_2)\right] + \cdots\cdots + \left[f(x_{n-1}) + f(x_n)\right]\right\}$$

Notice that in this formula all the function values appear twice except $f(x_0) = f(a)$ and $f(x_n) = f(b)$. So we can write

$$s = \frac{1}{2}h\left\{f(a) + 2f(x_1) + 2f(x_2) + \cdots\cdots 2f(x_{n-1}) + f(b)\right\}$$

The **trapezoidal method of integration** assumes that

$$\int_a^b f(x)dx \simeq s$$

As the number of subintervals increases and their width decreases the approximation is improved.

Example 8D

Use the trapezoidal method with four subintervals to evaluate $\int_0^1 e^x \, dx$.

Solution

If $a = 0$ and $b = 1$ and there are four subintervals then $h = 0.25$. The formula for the trapezoidal method gives

$$\int_0^1 e^x \, dx \simeq \frac{0.25}{2}\left\{e^0 + 2e^{0.25} + 2e^{0.5} + 2e^{0.75} + e^1\right\}$$
$$= 1.7272 \text{ to four decimal places.}$$

The exact solution is $e - 1 = 1.7183$ to four decimal places so that the error is 0.52%. The trapezoidal method with four intervals is correct to two significant figures.

The trapezoidal method is just one of many methods for approximating integrals. Another method is called **Simpson's rule** and is based on approximating the function by a set of quadratic polynomials instead of straight lines.

Consider an integral $\int_a^b f(x)dx$; divide the interval $[a,b]$ into an *even* number of subintervals and define a set of quadratic polynomials such that $p(x) = f(x)$ at the points x_{i-1}, x_i and x_{i+1}.

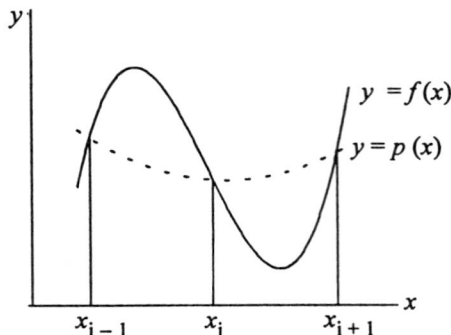

Figure 8.7

If the quadratic polynomial is $p(x) = ax^2 + bx + c$ then

$$ax_{i-1}^2 + bx_{i-1} + c = f(x_{i-1})$$
$$ax_i^2 + bx_i + c = f(x_i)$$
$$ax_{i+1}^2 + bx_{i+1} + c = f(x_{i+1})$$

Now if we integrate $p(x)$ over the subinterval $[x_{i-1}, x_{i+1}]$ and simplify the result we obtain

$$\int_{x_{i-1}}^{x_{i+1}} p(x)dx = \frac{a}{3}(x_{i+1} - x_{i-1})^3 + \frac{b}{2}(x_{i+1} - x_{i-1})^2 + c(x_{i+1} - x_{i-1})$$

Solving for the coefficients a, b and c and substituting gives

$$\int_{x_{i-1}}^{x_{i+1}} p(x)dx = \frac{h}{3}\left(f(x_{i-1}) + 4f(x_i) + f(x_{i+1})\right)$$

(The algebra involved is somewhat tedious so the result is stated and not derived).

Adding the contributions for each pair of subintervals gives

$$\int_a^b f(x)dx \cong \frac{h}{3}\left\{f(x_0) + 4f(x_1) + 2f(x_2) + 4f(x_3) + \cdots\cdots + 4f(x_{n-1}) + f(x_n)\right\}$$

where $f(x_0) = f(a)$ and $f(x_n) = f(b)$. This is called **Simpson's rule or method**. A convenient way of remembering this formula is

$$\int_a^b f(x)dx \cong \frac{h}{3} \text{ (first + last + 4 × odd values of } f + 2 \text{ × even values of } f).$$

Example 8E

Use Simpson's rule with four subintervals to evaluate $\int_0^1 e^x \, dx$.

Solution

With four subintervals we have $h = 0.25$. The four subintervals and corresponding values of the function $f = e^x$ are shown in Figure 8.8.

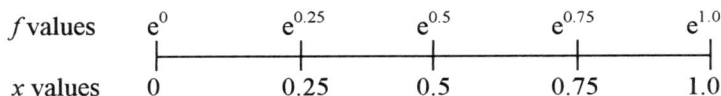

Figure 8.8

Simpson's rule with $h = 0.25$ gives

$$\int_0^1 e^x \, dx \cong \frac{0.25}{3}\left(e^0 + e^1 + 4\left(e^{0.25} + e^{0.75}\right) + 2e^{0.5}\right)$$
$$= 1.783 \text{ to four decimal places.}$$

Simpson's rule with four subintervals is now equal to the exact solution to four decimal places.

1.   Use the trapezoidal method and Simpson's rule to approximate the following integrals.

   (a) $\int_{-1}^{1} e^{x^2}\, dx$ with 8 subintervals

   (b) $\int_{0}^{1} \dfrac{2}{1+x^2}$ with 4 subintervals

   (c) $\int_{2}^{3} \dfrac{1}{x}\, dx$ with 6 subintervals

2.   Evaluate the integral

$$I = \int_{0.2}^{1} \frac{\sin(x)}{x}\, dx$$

   using Simpson's rule with (a) two, (b) four, and (c) six subintervals. Compare your results.

3.   Find an approximate value correct to three decimal places for the integral

$$I = \int_{0}^{\pi/2} \frac{x\sin(x)}{1+\cos^2(x)}\, dx$$

   using the trapezoidal method.

4.   A function f is given in tabular form as

x	0.8	1.0	1.2	1.4	1.6	1.8	2.0
$f(x)$	4.132	5.721	6.013	7.192	8.270	9.314	10.91

   Use (a) the trapezoidal method, and (b) Simpson's rule to approximate the integral $\int_{0.8}^{2.0} f(x)\, dx$ with $h = 0.2$.

9

Matrices

In this chapter we provide a short introduction to the algebra of matrices and their use in solving simultaneous linear equations.

9.1   INTRODUCTION

In mathematics we often represent quantities as arrays of numbers (or symbols). For example, the pair of numbers (2,3) could represent a point in the x-y plane of a cartesian coordinate system. In chapters 1 and 2 we introduced methods to fit a curve

to a set of data points displayed in the form $\begin{bmatrix} 1 & 7 \\ 2 & 11 \\ 3 & 15 \\ 4 & 19 \end{bmatrix}$. These arrays are examples of

matrices.

A **matrix** is an array of numbers arranged in regular rows and columns. Here are some examples of matrices.

$$\mathbf{A} = \begin{bmatrix} 1 & -3 \\ 0 & 4 \\ 5 & 1 \end{bmatrix} \qquad \text{is a matrix with 3 rows and 2 columns}$$

$$\mathbf{B} = \begin{bmatrix} -1 & 0 \\ 6 & 3 \end{bmatrix} \qquad \text{is a matrix with 2 rows and 2 columns}$$

$$\mathbf{C} = \begin{bmatrix} a_1 & a_2 & a_3 \\ b_1 & b_2 & b_3 \end{bmatrix} \qquad \text{is a matrix with 2 rows and 3 columns}$$

$$\mathbf{D} = \begin{bmatrix} 1 & 0 & 4 & -5 & 6 \end{bmatrix} \qquad \text{is a matrix with 1 row and 5 columns.}$$

In general, a matrix **A** with m rows and n columns is called an $m \times n$ **matrix**. A $1 \times n$ matrix is called a **row vector**; matrix **D** above is an example of a row vector.

An $m \times 1$ matrix is called a **column vector**; $\begin{bmatrix} 1 \\ 0 \\ 3 \end{bmatrix}$ is an example of a column vector.

Often these are just called **vectors**.

If the number of rows is equal to the number of columns then the matrix is called a **square matrix**. Matrix **B** above is an example of a square matrix.

A **diagonal matrix** is a square matrix with zeros everywhere except down the leading diagonal. For example, $\begin{bmatrix} 1 & 0 & 0 \\ 0 & 4 & 0 \\ 0 & 0 & -3 \end{bmatrix}$ is a 3×3 diagonal matrix. If the diagonal elements are all equal to 1 the matrix is called a **unit matrix**. The 4×4 unit matrix is

$$\begin{bmatrix} 1 & 0 & 0 & 0 \\ 0 & 1 & 0 & 0 \\ 0 & 0 & 1 & 0 \\ 0 & 0 & 0 & 1 \end{bmatrix}$$

Unit matrices are denoted by the letter **I**.

Matrices are often used to represent a set of simultaneous linear equations such as

$$\begin{aligned} 2x_1 - 4x_2 + x_3 &= 5 \\ x_1 + x_2 - 3x_3 &= -1 \\ 3x_1 - x_2 + x_3 &= 0 \end{aligned}$$

The properties of the matrix of coefficients $\begin{bmatrix} 2 & -4 & 1 \\ 1 & 1 & -3 \\ 3 & -1 & 1 \end{bmatrix}$ turn out to be very important in solving the equations. The unknown values of x can be written as a column vector $\mathbf{x} = \begin{bmatrix} x_1 \\ x_2 \\ x_3 \end{bmatrix}$. The three linear equations can be written as

$$\begin{bmatrix} 2 & -4 & 1 \\ 1 & 1 & -3 \\ 3 & -1 & 1 \end{bmatrix} \begin{bmatrix} x_1 \\ x_2 \\ x_3 \end{bmatrix} = \begin{bmatrix} 5 \\ -1 \\ 0 \end{bmatrix}.$$

If we denote the matrix of coefficients by **A** and the column vector on the right hand side by **b** then the set of linear equations can be written as

Ax = **b**

In general, to solve a set of simultaneous linear equations we need to solve matrix equations of the form **Ax** = **b** where **A** is a square matrix and **x** and **b** are column vectors.

Exercise 9A

Write each of the following sets of simultaneous linear equations in matrix form.

(a) $x - y = 4$
        $2x + 3y = 1$

(b) $-2x + y = -6$
        $x - y = 4$

(c) $3x - 6y + z = 7$
        $-2x + y - 3z = 2$
        $x + y + z = 0$

(d) $4x_1 - x_2 + x_3 = 1$
        $3x_1 + x_2 - 2x_3 = -3$
        $-x_1 - 4x_2 + x_3 = 5$

(e) $x_1 - 3x_2 + x_3 + 4x_4 - x_5 = 0$
        $-x_1 - 4.1x_3 + 2x_5 = 1.3$
        $0.3x_1 - 0.7x_2 + 4.1x_4 - x_5 = 2.7$
        $1.4x_1 - x_2 + 3.1x_4 = 0.4$
        $3x_1 + x_2 - x_3 + 2x_4 - x_5 = -3.5$

9.2   THE ARITHMETIC OF MATRICES

Matrices can be added, subtracted and multiplied together but you *cannot* divide by a matrix. Matrix division has no meaning.

Equality of matrices

Two matrices **A** and **B** are **equal** if they are the same size and the corresponding elements are equal. For example, if $\mathbf{A} = \begin{bmatrix} a_1 & a_2 & a_3 \\ b_1 & b_2 & b_3 \end{bmatrix}$ and $\mathbf{B} = \begin{bmatrix} 0 & -1 & 2 \\ 4 & 1 & -1 \end{bmatrix}$ then **A** = **B** means that

$$a_1 = 0 \qquad a_2 = -1 \qquad a_3 = 2$$

$$b_1 = 4 \qquad b_2 = 1 \qquad b_3 = -1$$

Addition and Subtraction of matrices

To add (or subtract) two matrices together they must be the same size. We then add (or subtract) corresponding elements. For example if

$$\mathbf{A} = \begin{bmatrix} 0 & -1 & 2 \\ 4 & 1 & -1 \end{bmatrix} \quad \text{and} \quad \mathbf{B} = \begin{bmatrix} 3 & 2 & -1 \\ 0 & -1 & -2 \end{bmatrix}$$

then we can add (or subtract) **A** and **B** because they are each 2×3 matrices to give

$$\mathbf{A} + \mathbf{B} = \begin{bmatrix} 0 & -1 & 2 \\ 4 & 1 & -1 \end{bmatrix} + \begin{bmatrix} 3 & 2 & -1 \\ 0 & -1 & -2 \end{bmatrix}$$

$$= \begin{bmatrix} 0+3 & -1+2 & 2-1 \\ 4+0 & 1-1 & -1-2 \end{bmatrix} = \begin{bmatrix} 3 & 1 & 1 \\ 4 & 0 & -3 \end{bmatrix}$$

$$\mathbf{A} - \mathbf{B} = \begin{bmatrix} 0 & -1 & 2 \\ 4 & 1 & -1 \end{bmatrix} - \begin{bmatrix} 3 & 2 & -1 \\ 0 & -1 & -2 \end{bmatrix}$$

$$= \begin{bmatrix} 0-3 & -1-2 & 2-(-1) \\ 4-0 & 1-(-1) & -1-(-2) \end{bmatrix} = \begin{bmatrix} -3 & -3 & 3 \\ 4 & 2 & 1 \end{bmatrix}$$

For matrix addition **A** + **B** = **B** + **A**. So we say that matrix addition is **commutative**. Note that **A** − **B** ≠ **B** − **A**. So we say matrix subtraction is not commutative.

Multiplication by a single number

To multiply a matrix by a single number we multiply each element by the number. For example, for **A** defined above

$$-2\mathbf{A} = -2\begin{bmatrix} 0 & -1 & 2 \\ 4 & 1 & -1 \end{bmatrix} = \begin{bmatrix} -2\times 0 & -2\times -1 & -2\times 2 \\ -2\times 4 & -2\times 1 & -2\times -1 \end{bmatrix}$$

$$= \begin{bmatrix} 0 & 2 & -4 \\ -8 & -2 & 2 \end{bmatrix}$$

Example 9A

Given the matrices $\mathbf{B} = \begin{bmatrix} 1 & 0 & 1 \\ -2 & 1 & 3 \end{bmatrix}$ and $\mathbf{C} = \begin{bmatrix} 0 & -1 & 5 \\ 4 & 0 & 1 \end{bmatrix}$, find the matrix **A** if

$$3\mathbf{A} + 2\mathbf{B} = \mathbf{C}$$

Solution

Applying the normal rules of algebra we subtract 2**B** from each side to give

$$3\mathbf{A} = \mathbf{C} - 2\mathbf{B}$$

and dividing by 3

$$\mathbf{A} = \tfrac{1}{3}(\mathbf{C} - 2\mathbf{B})$$

Substituting for **B** and **C**

$$\mathbf{A} = \frac{1}{3}\left(\begin{bmatrix} 0 & -1 & 5 \\ 4 & 0 & 1 \end{bmatrix} - 2\begin{bmatrix} 1 & 0 & 1 \\ -2 & 1 & 3 \end{bmatrix}\right)$$

$$= \frac{1}{3}\begin{bmatrix} -2 & -1 & 3 \\ 8 & -2 & -5 \end{bmatrix}$$

$$= \begin{bmatrix} -\frac{2}{3} & -\frac{1}{3} & 1 \\ \frac{8}{3} & -\frac{2}{3} & -\frac{5}{3} \end{bmatrix}$$

Multiplication of matrices

Two matrices **A** and **B** can be multiplied together to give a matrix **AB** provided the number of columns of **A** equals the number of rows of **B**. If this condition holds then **AB** is calculated by taking the products of every row of **A** with every column of **B** in a special way.

▩▩▩▩▩▩▩▩ **Example 9B** ▩▩▩

Let $\mathbf{A} = \begin{bmatrix} 2 & -1 \\ 1 & -3 \end{bmatrix}$, $\mathbf{B} = \begin{bmatrix} 4 & 1 & 5 \\ -3 & 0 & 7 \end{bmatrix}$ and $\mathbf{C} = \begin{bmatrix} 1 & 0 \\ -2 & 6 \\ 5 & -1 \end{bmatrix}$. Form all the possible matrix products of **A**, **B** and **C**.

Solution

We could attempt to form the products

$\qquad$ **AB, BA, AC, CA, BC** and **CB**

However, using the rule that the number of columns of the first matrix must equal the number of rows of the second matrix, the possible matrix products are

$\qquad$ **AB, CA** and **BC**

To calculate **AB**:

$$\mathbf{AB} = \begin{bmatrix} 2 & -1 \\ 1 & -3 \end{bmatrix} \begin{bmatrix} 4 & 1 & 5 \\ -3 & 0 & 7 \end{bmatrix}$$

$$= \begin{bmatrix} (2 \times 4 + -1 \times -3) & (2 \times 1 + -1 \times 0) & (2 \times 5 + -1 \times 7) \\ (1 \times 4 + -3 \times -3) & (1 \times 1 + -3 \times 0) & (1 \times 5 + -3 \times 7) \end{bmatrix} = \begin{bmatrix} 11 & 2 & 3 \\ 13 & 1 & -16 \end{bmatrix}$$

The method is that each number in the rth row of **A** is multiplied by each corresponding term in the sth column of **B** and added to give the element in the rth row, sth column of **AB**. For example, the element in the 2nd row and 3rd column of **AB** is shown by the following.

$$\begin{bmatrix} 2 & -1 \\ -1 & -3 \end{bmatrix} \begin{bmatrix} 4 & 1 & 5 \\ -3 & 0 & 7 \end{bmatrix} = \begin{bmatrix} * & * & * \\ * & * & c \end{bmatrix}$$

$c = (-1 \times 5) + (-3 \times 7)$

Similarly

$$\mathbf{CA} = \begin{bmatrix} 1 & 0 \\ -2 & 6 \\ 5 & -1 \end{bmatrix} \begin{bmatrix} 2 & -1 \\ 1 & -3 \end{bmatrix}$$

$$= \begin{bmatrix} (1 \times 2 + 0 \times -1) & (1 \times -1 + 0 \times -3) \\ (-2 \times 2 + 6 \times 1) & (-2 \times -1 + 6 \times -3) \\ (5 \times 2 + -1 \times 1) & (5 \times -1 + -1 \times -3) \end{bmatrix} = \begin{bmatrix} 2 & -1 \\ 2 & -16 \\ 9 & -2 \end{bmatrix}$$

$$\mathbf{BC} = \begin{bmatrix} 4 & 1 & 5 \\ -3 & 0 & 7 \end{bmatrix} \begin{bmatrix} 1 & 0 \\ -2 & 6 \\ 5 & -1 \end{bmatrix}$$

$$= \begin{bmatrix} (4 \times 1 + 1 \times -2 + 5 \times 5) & (4 \times 0 + 1 \times 6 + 5 \times -1) \\ (-3 \times 1 + 0 \times -2 + 7 \times 5) & (-3 \times 0 + 0 \times 6 + 7 \times -1) \end{bmatrix} = \begin{bmatrix} 27 & 1 \\ 32 & -7 \end{bmatrix}$$

The examples of matrix multiplication in Example 9B show several important properties:

1. In general **AB** ≠ **BA**. So we say that matrix multiplication is not commutative. For example,

$$\begin{bmatrix} 1 & 0 \\ 1 & 3 \end{bmatrix} \begin{bmatrix} 1 & 4 \\ -1 & 0 \end{bmatrix} = \begin{bmatrix} 1 & 4 \\ -2 & 4 \end{bmatrix}$$

$$\begin{bmatrix} 1 & 4 \\ -1 & 0 \end{bmatrix} \begin{bmatrix} 1 & 0 \\ 1 & 3 \end{bmatrix} = \begin{bmatrix} 5 & 12 \\ -1 & 0 \end{bmatrix}$$

2. If **A** is an $m \times r$ matrix and **B** is an $r \times n$ matrix then **AB** is an $m \times n$ matrix.

$$\begin{matrix} A & B & = & AB \\ m \times r & r \times n & & m \times n \end{matrix}$$

░░░░░░░░░░░░░░░ **Exercise 9B** ░░░░░░░░░░░░░░░

For the problems in this Exercise use the matrices given by

$$A = \begin{bmatrix} 0 & 1 \\ 1 & 0 \end{bmatrix} \qquad B = \begin{bmatrix} 1 & 3 & 0 \\ -1 & 2 & 1 \end{bmatrix} \qquad C = \begin{bmatrix} 1 & 4 & 2 \\ 0 & -2 & 1 \\ -1 & 1 & 0 \end{bmatrix}$$

$$D = \begin{bmatrix} 0 & 1 \\ 1 & 2 \\ -1 & -3 \end{bmatrix} \qquad E = \begin{bmatrix} 0 & 1 & 3 \\ -1 & 0 & 2 \\ 0 & 1 & 2 \end{bmatrix} \qquad F = \begin{bmatrix} 0 & 1 & -1 & 3 \\ -1 & 2 & 3 & 0 \\ 4 & 0 & -1 & 2 \end{bmatrix}$$

$$G = \begin{bmatrix} 1 & 0 \\ 0 & 1 \end{bmatrix}$$

and **I** is the unit matrix of appropriate size.

1.   Calculate each of the following (where possible).

       (a)   **C + E**   (b)   **A + B**   (c) $-3\mathbf{E}$

       (d)   **A + I**   (e)   **A − I**   (f) $2\mathbf{C} - 3\mathbf{E}$

       (g)   **E + F**   (h)   **AB**   (i)   **AC**

       (j)   **DE**   (k)   **ED**   (l)   **EF**

       (m)   **FC**   (n)   **BC**   (o)   **AI**

       (p)   **BI**   (q)   **IB**   (r)   **DI**

2.   If **x** is the column vector $\begin{bmatrix} x_1 \\ x_2 \\ x_3 \end{bmatrix}$ then expand **Bx, Cx** and **Ex**.

3.   Find the matrix **X** if

     (a)   **X** + **C** = **E**

     (b)   **X** + 3**C** = 5**E**

     (c)   2**X** – **C** = 2**E**

4.   If $\begin{bmatrix} a & 1 \\ 1 & b \end{bmatrix}$ **A** = **I**, find a and b.

5.   Show that if **X** is an $m \times m$ square matrix and if **I** is an $m \times m$ unit matrix then

$$XI = IX = X$$

9.3   TI-92 AND MATRICES

This section introduces the matrix algebra commands of the TI-92. Using your calculator for the arithmetic of matrices saves time in the same way as we use a calculator to manipulate numbers.

TI-92 ACTIVITY 9A

Clear the HOME screen and Data/Matrix Editor.

(A)   Consider the matrix $\mathbf{A} = \begin{bmatrix} 1 & 0 & 3 \\ -1 & 2 & 4 \end{bmatrix}$.

     To enter a matrix you need to open the Matrix Editor.

           Press **APPS**, select 6:Data/Matrix Editor, 3:New

     Respond to the next screen with Matrix, a name for the matrix in the variable box, Row dimension 2 and Col dimension 3. The screen dump in Figure 9.1 shows this set of instructions.

     Using the cursor pad highlight the relevant box and input the matrix.

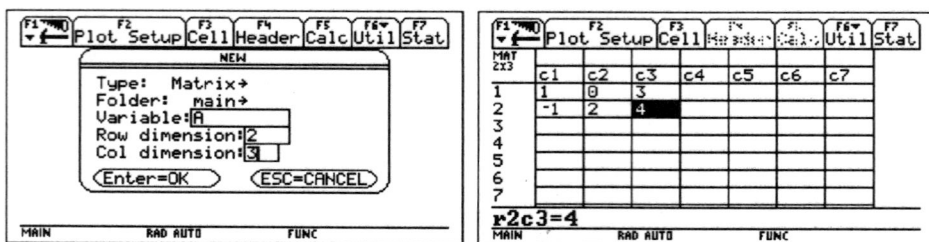

Figure 9.1   Entering Matrix **A**

(B)   Now enter the following matrices in the same way:

$$\mathbf{B} = \begin{bmatrix} 0 & -2 & 1 \\ -1 & 1 & 3 \end{bmatrix} \qquad \mathbf{C} = \begin{bmatrix} 0 & -1 \\ 2 & 3 \\ -4 & 5 \end{bmatrix}$$

and label them as **B** and **C**.

(C)   Return to the HOME screen. Type A + B **ENTER**. You should get the result

$$\mathbf{A} + \mathbf{B} = \begin{bmatrix} 1 & -2 & 4 \\ -2 & 3 & 7 \end{bmatrix}$$

Type B + C **ENTER**.
Explain the response of the TI-92.

(D)   To multiply two matrices together you use the multiplication symbol ×. Type
A × C **ENTER**. Figure 9.2 shows the outcome of this activity. Notice that on
the HOME screen a dot appears between A and B.

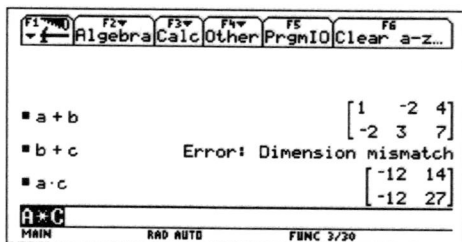

Figure 9.2   Multiplication of matrices

(E)   What happens if you try AB? Why will the TI-92 not simplify the product?

(F)   An alternative method of entering a matrix on the TI-92 is to use the square brackets [ ]. For example, to input $\mathbf{M} = \begin{bmatrix} 1 & 0 & 3 \\ -1 & 2 & 4 \end{bmatrix}$ type

>   [1, 0, 3; –1, 2, 4] **STO▷ M ENTER**

Figure 9.3 shows the HOME screen for this activity.

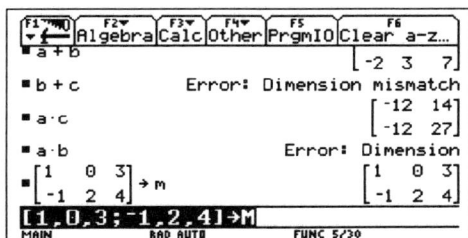

Figure 9.3   Entering a matrix directly

(G)   Use the TI-92 to check all your answers to problems 1 and 3 of Exercise 9B.

9.4   SQUARE MATRICES AND THEIR INVERSES

In section 9.3 we said that division of matrices is not defined but an operation similar to division is possible if we use the **inverse** of a matrix.

Suppose that **A** and **B** are two $n \times n$ square matrices and that **I** is the associated unit matrix. If

>   **AB** = **I**

then the matrix **B** is called the **inverse matrix** of **A** and is written as $\mathbf{A}^{-1}$ so that

> $\mathbf{AA}^{-1} = \mathbf{I}$ and $\mathbf{A}^{-1}\mathbf{A} = \mathbf{I}$

░░░░░░░░░░░░ **Example 9C** ░░░░░░░░░░░░░░░░░░░░░░░░░░░░░░░░░░

Given that $\mathbf{A} = \begin{bmatrix} a & b \\ c & d \end{bmatrix}$ and $\mathbf{B} = \dfrac{1}{(ad - bc)}\begin{bmatrix} d & -b \\ -c & a \end{bmatrix}$ where $ad - bc \neq 0$ show that

$$\mathbf{AB} = \mathbf{BA} = \mathbf{I}$$

Solution

$$\mathbf{AB} = \frac{1}{(ad - bc)}\begin{bmatrix} a & b \\ c & d \end{bmatrix} \cdot \begin{bmatrix} d & -b \\ -c & a \end{bmatrix}$$

$$= \frac{1}{(ad - bc)}\begin{bmatrix} ad - bc & 0 \\ 0 & -bc + ad \end{bmatrix}$$

$$= \begin{bmatrix} 1 & 0 \\ 0 & 1 \end{bmatrix} = \mathbf{I}$$

$$\mathbf{BA} = \frac{1}{(ad - bc)}\begin{bmatrix} d & -b \\ -c & a \end{bmatrix}\begin{bmatrix} a & b \\ c & d \end{bmatrix}$$

$$= \frac{1}{(ad - bc)}\begin{bmatrix} ad - bc & 0 \\ 0 & -cb + ad \end{bmatrix}$$

$$= \begin{bmatrix} 1 & 0 \\ 0 & 1 \end{bmatrix} = \mathbf{I}$$

This example provides a simple formula for finding the inverse of a 2×2 matrix. However, it also shows that the inverse of a matrix may not always exist. It can be seen that if $ad - bc$ is zero then the matrix **B** has no meaning. The quantity $(ad-bc)$ associated with the matrix **A** is called the **determinant** of A. Every square matrix has associated with it a number called its determinant. We will not develop the theory of determinants in this text but one important consequence of the determinant of a matrix is associated with the inverse matrix.

If the determinant associated with a matrix **A** is zero then the matrix is said to be **singular** and its inverse $\mathbf{A}^{-1}$ does not exist.

Example 9D

Show that the matrix $\mathbf{A} = \begin{bmatrix} 1 & 2 \\ 2 & 4 \end{bmatrix}$ does not have an inverse.

Solution

The determinant associated with this matrix is $(1 \times 4 - 2 \times 2) = 0$ so that

$$\mathbf{B} = \frac{1}{(ad - bc)} \begin{bmatrix} d & -b \\ -c & a \end{bmatrix} \text{ has no meaning.}$$

One use of the inverse matrix is in the solution of simultaneous linear equations. Suppose that we are given a set of n simultaneous equations in the matrix form

$$\mathbf{Ax} = \mathbf{b}$$

where **x** and **b** are $n \times 1$ column vectors and **A** is an $n \times n$ matrix.

If we multiply both sides of the equation on the left by $\mathbf{A}^{-1}$ we obtain

$$\mathbf{A}^{-1}\mathbf{Ax} = \mathbf{A}^{-1}\mathbf{b}$$
$$\mathbf{Ix} = \mathbf{A}^{-1}\mathbf{b}$$
$$\mathbf{x} = \mathbf{A}^{-1}\mathbf{b}$$

Hence if we can find $\mathbf{A}^{-1}$ we can solve for **x**.

Solve the linear equations

$$2x + y = 1$$
$$x + 4y = 11$$

Solution

In matrix form the problem can be written as

$$\begin{pmatrix} 2 & 1 \\ 1 & 4 \end{pmatrix} \begin{pmatrix} x \\ y \end{pmatrix} = \begin{pmatrix} 1 \\ 11 \end{pmatrix}$$

so that

$$\begin{pmatrix} x \\ y \end{pmatrix} = \begin{pmatrix} 2 & 1 \\ 1 & 4 \end{pmatrix}^{-1} \begin{pmatrix} 1 \\ 11 \end{pmatrix}$$

Using the formula of Example 9C gives

$$\begin{pmatrix} 2 & 1 \\ 1 & 4 \end{pmatrix}^{-1} = \frac{1}{7} \begin{pmatrix} 4 & -1 \\ -1 & 2 \end{pmatrix}$$

and then

$$\begin{pmatrix} x \\ y \end{pmatrix} = \frac{1}{7} \begin{pmatrix} 4 & -1 \\ -1 & 2 \end{pmatrix} \begin{pmatrix} 1 \\ 11 \end{pmatrix} = \frac{1}{7} \begin{pmatrix} -7 \\ 21 \end{pmatrix} = \begin{pmatrix} -1 \\ 3 \end{pmatrix}$$

Hence $x = -1$ and $y = 3$.

Exercise 9C

1. If possible find the inverse of each of the following matrices:

(a) $A = \begin{pmatrix} 1 & 5 \\ 3 & -2 \end{pmatrix}$

(b) $B = \begin{pmatrix} 0 & 2 \\ -1 & 4 \end{pmatrix}$

(c) $C = \begin{pmatrix} -1 & 2 \\ 6 & -12 \end{pmatrix}$

(d) $D = \begin{pmatrix} 0.3 & -0.4 \\ 1 & 0.7 \end{pmatrix}$

2. Use the inverse of a matrix to solve the following equations.

(a) $\begin{aligned} x + 5y &= 2 \\ 3x - 2y &= 1 \end{aligned}$

(b) $\begin{aligned} 0.3x - 0.4y &= 1.2 \\ x + 0.7y &= 0.8 \end{aligned}$

(c) $\begin{aligned} 2x + 5y &= 19 \\ 3x + y &= 9 \end{aligned}$

(d) $\begin{aligned} ax + by &= r \\ cx + dy &= s \end{aligned}$

3. Given that $A = \begin{bmatrix} 1 & -1 & 2 \\ 2 & 0 & 1 \\ -1 & 3 & 4 \end{bmatrix}$ and $B = \dfrac{1}{18}\begin{pmatrix} -3 & 10 & -1 \\ -9 & 6 & 3 \\ 6 & -2 & 2 \end{pmatrix}$ show that $B = A^{-1}$

and hence solve the linear equations

$$\begin{aligned} x - y + 2z &= 2 \\ 2x \qquad + z &= -1 \\ -x + 3y + 4z &= 3 \end{aligned}$$

4. Given that $A = \begin{bmatrix} 0 & 1 & -1 & 2 \\ -1 & 3 & 1 & 1 \\ 4 & -2 & 0 & -1 \\ 1 & 1 & -1 & 2 \end{bmatrix}$ and $B = \begin{pmatrix} -1 & 0 & 0 & 1 \\ -6 & -0.5 & -1.5 & 5.5 \\ 9 & 1.5 & 2.5 & -8.5 \\ 8 & 1 & 2 & -7 \end{pmatrix}$ show

that $A = B^{-1}$ and hence solve the linear equations

$$\begin{aligned} -x_1 \qquad\qquad + x_4 &= 2 \\ -6x_1 - 0.5x_2 - 1.5x_3 + 5.5x_4 &= 0 \\ 9x_1 + 1.5x_2 + 2.5x_3 - 8.5x_4 &= 0.1 \\ 8x_1 + x_2 + 2x_3 - 7x_4 &= 0 \end{aligned}$$

5.   The TI-92 can be used to find matrix inverses and determinants.

(A)   Enter the matrix $\mathbf{A} = \begin{pmatrix} 1 & -1 & 2 \\ 2 & 0 & 1 \\ -1 & 3 & 4 \end{pmatrix}$.

Now type A $\wedge(-1)$ **ENTER**.
The HOME screen shows the inverse of the matrix **A**.

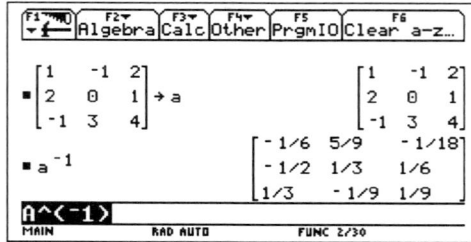

Figure 9.4   The inverse of **A**

(B)   Use the TI-92 to find the inverse of each of the following matrices.

(i) $\begin{pmatrix} 2 & 1 \\ -1 & 4 \end{pmatrix}$ (ii) $\begin{pmatrix} 1 & 2 & -1 \\ -1 & 1 & 2 \\ 2 & -1 & 1 \end{pmatrix}$

(iii) $\begin{pmatrix} 1 & 1 & 2 & 3 \\ 1 & 2 & 4 & -1 \\ 2 & 4 & -1 & 1 \\ 0 & 1 & 2 & -1 \end{pmatrix}$ (iv) $\begin{pmatrix} 1 & 2 & 3 \\ 4 & 5 & 6 \\ 7 & 8 & 9 \end{pmatrix}$

(C)   The TI-92 can be used to find the determinant associated with a matrix. Enter

the matrix $\begin{pmatrix} 1 & -1 & 2 \\ 2 & 0 & 1 \\ -1 & 3 & 4 \end{pmatrix}$ and label it as A.

Now type DET(A) **ENTER**. The TI-92 gives the answer 18.
Use the TI-92 to find the determinants associated with the matrices in (B).
What can you say about the inverse of the matrix in (iv)?

(D) Given the set of linear equations

$$2x_1 + x_2 + 3x_3 = 5$$
$$2x_2 + x_3 = 4$$
$$3x_1 + x_2 + 6x_3 = 10$$

(i) Write the equations in matrix form $Ax = b$;

(ii) Find the inverse of A;

(iii) Hence solve the equations for x_1, x_2 and x_3.

(E) Use the steps in part (D) to solve the following:

(i) $$x_1 + 2x_2 - x_3 = 1$$
$$-x_1 + x_2 + 2x_3 = -5$$
$$2x_1 - x_2 + x_3 = 4$$

(ii) $$x + 2y + 3z + w = 4$$
$$2x + y + z + w = 3$$
$$x + 3y + z + w = 2$$
$$y + z + 2w = 1$$

(iii) $$x_1 + 2x_2 + 3x_3 + x_4 = 5$$
$$2x_1 + x_2 + x_3 + x_4 = 3$$
$$x_1 + 2x_2 + x_3 = 4$$
$$x_2 + x_3 + 2x_4 = 0$$

(iv) $$-x_1 + x_4 = 2$$
$$-6x_1 - 0.5x_2 - 1.5x_3 + 5.5x_4 = 0$$
$$9x_1 + 1.5x_2 + 2.5x_3 - 8.5x_4 = 0.1$$
$$8x_1 + x_2 + 2x_3 - 7x_4 = 0$$

6.   The TI-92 can solve systems of simultaneous linear equations directly. Consider the equations

$$x + 2y = 3$$
$$-x - 3y = 4$$

Type simult([1,2;–1,–3],[3;4]) **ENTER**

The TI-92 responds with the column vector $\begin{bmatrix} 17 \\ -7 \end{bmatrix}$. So $x = 17$ and $y = -7$ is the solution of the equations.

The simult utility needs the following parameters

simult(matrix, vector)

The matrix is the matrix of coefficients in the equation $\mathbf{Ax} = \mathbf{b}$ and the vector is the right hand side, $\mathbf{b}$.

The screen dump in Figure 9.5 shows the solution of the set of linear equations.

$$2x_1 + x_2 + 3x_3 = 5$$
$$2x_2 + x_3 = 4$$
$$3x_1 + x_2 + 6x_3 = 10$$

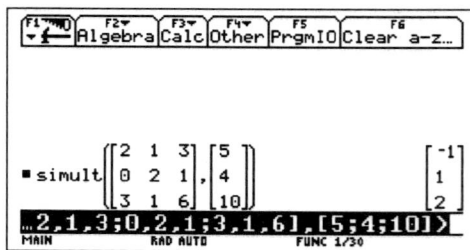

Figure 9.5   Solving simultaneous equations directly

Use the simult utility to solve the sets of linear equations in problem 5 (E).

10

Complex numbers

10.1 THE OCCURRENCE OF COMPLEX NUMBERS

Complex numbers often arise when solving polynomial equations. For example, if we use the TI-92 to solve the following equation

$$x^4 + x^3 - x^2 + x - 2 = 0$$

we obtain two solutions, $x = 1$ and $x = -2$. Since we have a polynomial of degree 4 we would expect four solutions. A graph of the function shows only two roots (see Figure 10.1).

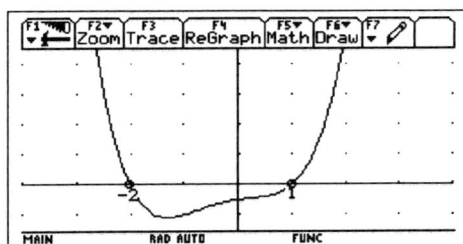

Figure 10.1   Solving $x^4 + x^3 - x^2 + x - 2 = 0$

We can explore this polynomial further. Since $x = 1$ and $x = -2$ are solutions we can deduce that $(x - 1)$ and $(x + 2)$ are factors so the polynomial can be written as

$$(x - 1)(x + 2)(x^2 + 1) = 0$$

Now we can see that the two missing roots are solutions of the quadratic equation

$$x^2 + 1 = 0$$
$$x = \pm\sqrt{-1}$$

If you try $\sqrt{(-1)}$ on your TI-92 then the response is 'Non-real result'. The two solutions $\sqrt{-1}$ and $-\sqrt{-1}$ are examples of a new set of numbers called **complex numbers**. Because of the importance of complex numbers the special symbol i is reserved for $\sqrt{-1}$. We write

$$i = \sqrt{-1}$$

You will find this notation on the TI-92 using **2nd I**. Use your TI-92 to find the following powers of i:

$$i^2, \, i^3, \, i^4, \, i^5, \, \frac{1}{i}, \, \frac{1}{i^2}$$

The screen dump in Figure 10.2 shows some of the answers. The following example proves some of these results.

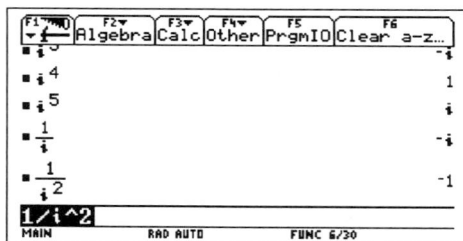

Figure 10.2   Evaluating powers of $i = \sqrt{-1}$

Example 10A

Use the definition of i to prove that $i^5 = i$ and $\dfrac{1}{i} = -i$.

Solution

We use the basic result that $i = \sqrt{-1}$ so $i^2 = -1$.

$$i^5 = i^2 \times i^2 \times i = (-1) \times (-1) \times i = i$$

$$\frac{1}{i} = \frac{i}{i \times i} = \frac{i}{i^2} = \frac{i}{(-1)} = -i$$

The introduction of the complex numbers i and $-i$ suggests that the polynomial equation

$$x^4 + x^3 - x^2 + x - 2 = 0$$

does indeed have four solutions, $x = 1$, $x = -2$, $x = i$ and $x = -i$.
The TI-92 will solve the equation completely but we need to alert the calculator that complex solutions are expected.

Press **F2** select A:Complex,1:cSolve(

and type

$$x^4 + x^3 - x^2 + x - 2 = 0,x) \ \textbf{ENTER}$$

This gives the four solutions, two real and two complex.

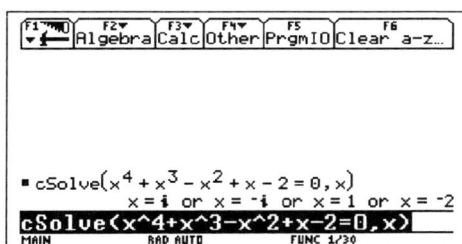

Figure 10.3   Using cSolve to find all the solutions

TI-92 ACTIVITY 10A

Use the cSolve command on the TI-92 to find the solution of the following equations.

$$x^2 + 2 = 0$$
$$x^2 + x + 3 = 0$$
$$x^3 + 15x - 4 = 0$$
$$x^4 - 3x^2 + 2x - 4 = 0$$
$$x^4 - x^3 + 2x^2 + 2x - 4 = 0$$
$$x^6 + x^5 - x^3 - 11x^2 + 2x - 12 = 0$$

What is the relationship between the number of solutions and the degree of the polynomial?

Do you notice a pattern in the complex numbers?

Investigate your conjectures by solving some polynomial equations of your choice.

Summary

You have probably deduced that the number of solutions equals the degree of the polynomial and that the complex numbers appear to be formed in pairs that are identical except for a sign change between the first and second terms. For example when solving $x^4 - x^3 + 2x^2 + 2x - 4 = 0$ we had the pair of complex numbers

$$0.589754 + 1.74454i \quad \text{and} \quad 0.589754 - 1.74454i$$

We see again the meaning of i from the example

$$x^2 + 2 = 0$$

Rearrange this equation to give

$$x^2 = -2$$

and take square roots

$$x = \pm\sqrt{-2}$$

which the TI-92 has written as $\pm 1.4142i \ (= \pm\sqrt{2}i)$.

     All the solutions in the TI-92 activity were of the form $z = a + bi$. If $b = 0$ the number is a familiar one called **a real number**. If $b \neq 0$ then the number is a **complex number**.

Here a and b are real numbers; a is called the **real part** of the complex number z and written as Re(z); and b is called the **imaginary part** of the complex number z and written as Im(z).

Associated with a complex number $z = a + bi$ is the complex number $z* = a - bi$ which is called the **complex conjugate**. You will have noticed how a complex number and its conjugate appear as pairs of solutions of polynomial equations.

Example 10B

Write down the real and imaginary parts and the complex conjugate of each of the following complex numbers.

(a) $-4 + 2i$
(b) $3 - 6i$
(c) $1.72 - 4.21i$

Solution

Table 10.1 shows the solutions.

Table 10.1

	complex number	real part	imaginary part	complex conjugate
(a)	$-4 + 2i$	-4	2	$-4 - 2i$
(b)	$3 - 6i$	3	-6	$3 + 6i$
(c)	$1.72 - 4.21i$	1.72	-4.21	$1.72 + 4.21i$

Example 10C

Find the quadratic equation whose solutions are the complex numbers $3 + 4i$ and $3 - 4i$.

Solution

If a and b are solutions of a quadratic equation then

$$(x - a)(x - b) = 0$$

or

$$x^2 - (a + b)x + ab = 0$$

If $a = 3 + 4i$ and $b = 3 - 4i$ then $a + b = 6$ and $ab = 3^2 + 4^2 = 25$. So the quadratic equation is $x^2 - 6x + 25 = 0$.

Exercise 10A

1.   Write down the real and imaginary parts and the complex conjugate of each of the following complex numbers.

    (a) $2 - 3i$,
    (b) $6 + 2i$,
    (c) $1.73 - 2.19i$,
    (d) $1.7 + 4.6i$,
    (e) $-5.17 + 1.03i$,
    (f) $-4i$,
    (g) $+17i$,
    (h) $x - yi$,
    (i) $-p + qi$

2.   Solve these equations.

    (a) $x^2 + 2x + 3 = 0$
    (b) $x^2 + 9 = 0$
    (c) $2x^2 - x + 1 = 0$
    (d) $3x^2 + 2x + 2 = 0$

3.   Show that for any complex number $z = a + bi$,

    (a) $zz* = a^2 + b^2$
    (b) $z + z* = 2\mathrm{Re}(z)$
    (c) $z - z* = 2\mathrm{Im}(z)$

where $z* = a - bi$.

4.   Find the quadratic equation that each pair of complex numbers satisfies.

    (a) $2 - 3i,\ 2 + 3i$
    (b) $0.2 + 0.5i,\ 0.2 - 0.5i$
    (c) $3 - 5i,\ 3 + 5i$
    (d) $a + bi,\ a - bi$

10.2   GRAPHICAL REPRESENTATION OF COMPLEX NUMBERS

The complex number $z = a + bi$ can be represented on a diagram using cartesian coordinates by the point (a,b). The line from the origin to (a,b) is often drawn to help visualise the number. For example, Figure 10.4 shows the complex numbers $3 + 2i$, $-1 + i$, $-2 - 3i$ and $4 - 2i$.

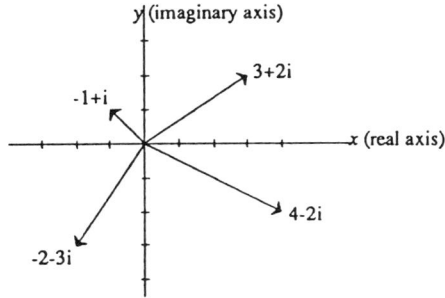

Figure 10.4   The Argand Diagram

The representation of a complex number in this way is called an **Argand Diagram**.

From the Argand diagram we can associate the size of a complex number with the length of the line and the inclination of a complex number with the angle the line makes with the x-axis. The size and inclination of a complex number are called the **modulus** and **argument** respectively.

For the complex number $z = a + bi$

modulus $= |z| = r = \sqrt{a^2 + b^2}$

argument $= \arg(z) = \theta$

        $=$ angle of OP to positive x direction.

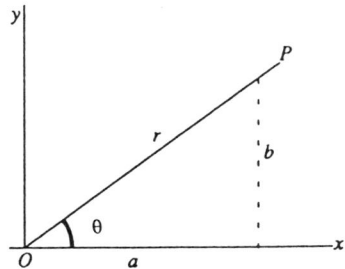

Figure 10.5

If z lies above the x-axis $0 \le \arg(z) \le \pi$ whereas if z lies below the x-axis $-\pi < \arg(z) < 0$.

Show the following complex numbers on an Argand diagram and find the modulus
and argument of each one.

$$3 + 2i, \quad -4 - 2i, \quad 1 - 2i$$

Solution

Figure 10.6 shows the complex numbers on an Argand diagram.

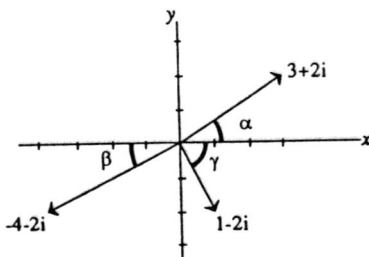

Figure 10.6

For 3 + 2i: $\left|3+2i\right| = \sqrt{3^2 + 2^2} = \sqrt{13}$

$$\arg(3+2i) = \alpha = \arctan\frac{2}{3} = 0.522$$

For –4 – 2i: $\left|-4-2i\right| = \sqrt{4^2 + 2^2} = \sqrt{20}$

$$\arg(-4-2i) = -(\pi - \beta) = -\left(\pi - \arctan\left(\frac{2}{4}\right)\right) = -2.68$$

For 1 – 2i: $\left|1-2i\right| = \sqrt{1^2 + 2^2} = \sqrt{5}$

$$\arg(1-2i) = -\gamma = -\arctan\left(\frac{2}{1}\right) = -1.11$$

Figure 10.6 shows that the relationships between a,b and r,θ are

$a = r\cos\theta$ and $b = r\sin\theta$.

Hence the complex number $z = a + bi$ can be written in terms of r and θ as

$$z = r\cos\theta + ir\sin\theta = r(\cos\theta + i\sin\theta)$$

This is called **the polar form** of the complex number.

For example for the complex number $1 - 2i$ of Example 10D the polar form is written as $\sqrt{5}(\cos 1.11 - i\sin 1.11)$ since $\cos(-1.11) = \cos(1.11)$ and $\sin(-1.11) = -\sin(1.11)$.
     The TI-92 gives the modulus (using abs(z) for absolute value) and argument (using angle(z)) of a complex number directly. Consider the complex number $1 - 2i$ of Example 10D. The TI-92 gives

$$\text{abs}(1 - 2i) = \sqrt{5}$$

$\text{angle}(1 - 2i) = -1.10715$ in RAD MODE or -63.4349 in DEG MODE.

Figure 10.7 shows the screens for these calculations.

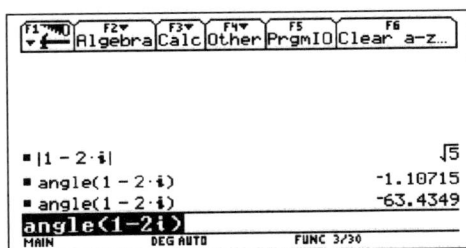

Figure 10.7 Using the TI-92 to find the modulus (abs) and
argument (angle) of a complex number

1.   Determine the modulus and argument of the following complex numbers. Show each of them on an Argand diagram. Write each of them in polar form.

(a) i (b) $1 + i$ (c) $2 - 3i$

(d) $-1 - 4i$ (e) $-3 - 4i$ (f) $-5 + 12i$

(g) 1 (h) $-3i$ (i) -4

(j) $1 - \sqrt{3}i$ (k) $-1 - \sqrt{3}i$ (l) $1.32 - 6.21i$

Check your answers using the TI-92.

2.   Find the solutions of the following equations and write them in polar form.

(a) $x^2 + 2x + 2 = 0$

(b) $x^3 + 8 = 0$

3.   Use the TI-92 to evaluate the following.

(a) $|3 + 4i|$ (b) $\arg(3 - 4i)$ (c) $|(1+i)(3-2i)|$

(d) $\left|\dfrac{5-12i}{1+i}\right|$ (e) $\arg\left(\dfrac{5-12i}{1+i}\right)$ (f) $|(3-i)^2(2+i)^3|$

10.3   THE ARITHMETIC OF COMPLEX NUMBERS

The addition, subtraction and multiplication of complex numbers is straightforward. For example,

$(2-3i) + (1+4i) = 2 - 3i + 1 + 4i = (2+1) + (-3i+4i) = 3 + i$

$(2-3i) - (1+4i) = 2 - 3i - 1 - 4i = (2-1) + (-3i-4i) = 1 - 7i$

$(2-3i)(1+4i) = 2 + 8i - 3i - 12i^2 = 2 + 8i - 3i + 12 = 14 + 5i$

The division of complex numbers uses the complex conjugate. Consider $z = \dfrac{(2-3i)}{(1+4i)}$.

We multiply top and bottom by $(1-4i)$ giving

$$z = \frac{(2-3i)(1-4i)}{(1+4i)(1-4i)} = \frac{2-8i-3i+12i^2}{1-4i+4i-16i^2} = \frac{2-8i-3i-12}{1-4i+4i+16} = \frac{-10-11i}{17}$$

Exercise 10C

1. For each pair of complex numbers find

$$z_1 + z_2, \quad z_1 - z_2, \quad z_1z_2, \quad \frac{z_1}{z_2}, \quad z_1^2 \quad \text{and} \quad 3z_1 - 2z_2.$$

   (a) $z_1 = i, z_2 = 3 + 4i$

   (b) $z_1 = -3 + 4i, z_2 = 1 - i$

   (c) $z_1 = 1 + i, z_2 = 1 - i$

   (d) $z_1 = -6 - i, z_2 = 1 - 2i$

   (e) $z_1 = 3 + 2i, z_2 = 5 - i$

   (f) $z_1 = -3 - 4i, z_2 = 5 + 12i$

   Check your answers using the TI-92.

2. With $z = 3 - 2i$ find

   (a) iz (b) $z*$ (c) $\dfrac{1}{z}$ (d) $\dfrac{1}{z*}$.

3. Show on an Argand diagram that the effect of multiplying a complex number by i is a rotation of $\dfrac{\pi}{2}$ (ie. 90°) anticlockwise.

4. Find the values of a and b if

   (a) $a + bi = \dfrac{3}{\cos(\theta) + i\sin(\theta)}$

   (b) $a + bi = \dfrac{2 - i}{1 + \cos(\theta) - i\sin(\theta)}$

5. Find the real and imaginary parts of z when

$$\frac{1}{z} = \frac{1}{1+i} - \frac{1}{2-i}$$

6. With the TI-92 use the angle command to:

   (a) Show that the argument of $(1+i)^3$ is three times the argument of $(1 + i)$;

   (b) Show that the argument of $\left(-\frac{1}{2}+\frac{\sqrt{3}}{2}i\right)^4$ is four times the argument of $\left(-\frac{1}{2}+\frac{\sqrt{3}}{2}i\right)$;

   Propose a rule for the modulus and argument of z^n in terms of n and the modulus and argument of z. Use the TI-92 to test your rule for i^6, $(2+3i)^3$, $(4-2i)^5$, $(1-i)^7$, $(4.7+1.7i)^9$.

   Hence write z^n in polar form given that $z = r(\cos\theta + i\sin\theta)$.

7. (a) Use the TI-92 to simplify the expression e^{3i} in APPROX mode.

   (b) Find the modulus and argument of the answer to part (a).

   Repeat parts (a) and (b) for the expressions e^{4i}, e^{-i} and $e^{0.4i}$.

   What can you deduce about $\arg(e^{bi})$?

10.4 EULER'S FORMULA

We now bring together the results of problems 6 and 7 of Exercise 10C.

In problem 6 you may have deduced that if

$$z = r(\cos\theta + i\sin\theta)$$

then z^n can be written as

$$z^n = r^n(\cos(n\theta) + i\sin(n\theta))$$

This says that the argument of z^n is n times the argument of z. This important result is known as **de Moivre's Theorem**.

In problem 7 you will have seen that e^{bi} is a complex number with modulus 1 and argument b. So e^{bi} can be written as

$$e^{bi} = \cos b + i\sin b$$

This is known as **Euler's formula**. It provides a fascinating and powerful link between the exponential and trigonometric functions.

We can now write a complex number in three different ways

$z = x + iy$ **cartesian form**

$z = r(\cos\theta + i\sin\theta)$ **polar form**

$z = re^{i\theta}$ **exponential form**.

Example 10E

Use Euler's formula to prove de Moivre's Theorem.

Solution

Let $z = r(\cos\theta + i\sin\theta)$. Then using Euler's formula we have

$$z = re^{i\theta}$$

Using the rules of indices, we have, for any n

$$z^n = \left(re^{i\theta}\right)^n = r^n e^{i(n\theta)}$$

Using Euler's formula again

$$z^n = r^n(\cos n\theta + i\sin n\theta)$$

░░░░░░░░░░░░░░░░░░░░ **Example 10F** ░░░░░░░░░░░░░░░░░░░░░░░░░░░░░░

Write $-1 + i$ in the polar form and hence evaluate $(-1+i)^8$.

Solution

$$|-1+i| = \sqrt{2} \quad \text{and} \quad \arg(-1+i) = \frac{3\pi}{4}$$

so that

$$-1+i = \sqrt{2}\left[\cos(3\pi/4) + i\sin(3\pi/4)\right]$$

Now

$$(-1+i)^8 = (\sqrt{2})^8\left[\cos(3\pi/4) + i\sin(3\pi/4)\right]^8$$

$$= 2^4\left[\cos\left(8 \times \frac{3\pi}{4}\right) + i\sin\left(8 \times \frac{3\pi}{4}\right)\right]$$

$$= 2^4\left[\cos(6\pi) + i\sin(6\pi)\right] = 2^4[1+0i]$$

$$= 16$$

░░░░░░░░░░░░░░░░░░░░ **Exercise 10D** ░░░░░░░░░░░░░░░░░░░░░░░░░░░░

1.   Write the following complex numbers in polar and exponential form.

(a) $3 + 4i$ (b) $1 - i$ (c) $-\dfrac{\sqrt{3}}{2} - \dfrac{1}{2}i$

(d) $5 - 12i$ (e) $-2 + 2i$ (f) $-0.3 - 0.7i$

(g) $\dfrac{1}{3+4i}$ (h) $\dfrac{1}{1+2i}$ (i) $\dfrac{1-i}{5-12i}$

2.   Write the following complex numbers in cartesian form.

(a) $e^{i\pi}$ (b) $3e^{-i\pi/2}$ (c) $e^{-i\pi}$

(d) $2e^{0.7i}$ (e) $4e^{-0.1i}$ (f) $0.9e^{i\pi/4}$

Show each complex number on an Argand diagram.

3.   Given $z_1 = 2e^{i\pi/4}$ and $z_2 = 3e^{-i\pi/2}$ find the modulus and arguments of

(a) z_1^2 (b) z_2^3 (c) $z_1^2 z_2^3$ (d) $\dfrac{z_1^2}{z_1^3}$

4.   Repeat problem 3 for $z_1 = 4e^{0.7i}$ and $z_2 = 0.5e^{-0.1i}$.

5.   Write $3 - 4i$ in polar form and hence evaluate $(3-4i)^6$.

6.   Write $1 + i$ in polar form and hence evaluate $(1+i)^{12}$.

7.   By expanding $(\cos\theta + i\sin\theta)^3$ and using de Moivre's theorem, write $\cos3\theta$ and $\sin3\theta$ in terms of $\cos\theta$ and $\sin\theta$.

10.5   FINDING ROOTS OF COMPLEX NUMBERS

One of the most common uses of de Moivre's theorem is to find the roots of a complex number such as $\sqrt{z}$ and $\sqrt[4]{z}$. The following examples show the method of approach.

Example 10G

Find the three cube roots of $1 + i$.

Solution

The modulus of $1 + i$ is $\sqrt{2}$ and the argument of $1 + i$ is $\dfrac{\pi}{4}$, so that

$$1+i = \sqrt{2}\left[\cos\left(\frac{\pi}{4}\right) + i\sin\left(\frac{\pi}{4}\right)\right]$$

Take the cube root of each side

$$(1+i)^{\frac{1}{3}} = (\sqrt{2})^{\frac{1}{3}}\left[\cos\left(\frac{\pi}{4}\right) + i\sin\left(\frac{\pi}{4}\right)\right]^{\frac{1}{3}}$$

$$= 2^{\frac{1}{6}}\left[\cos\left(\frac{\pi}{12}\right) + i\sin\left(\frac{\pi}{12}\right)\right]$$

by de Moivre's theorem. In cartesian form one cube root of $(1+i)$ is $1.084 + 0.291i$.

To find the other two cube roots we need to observe that the argument of a complex number is not unique. Because of the periodicity of $\cos\theta$ and $\sin\theta$ we could write

$$z = r(\cos\theta + i\sin\theta) = r\left[\cos(\theta + 2\pi k) + i\sin(\theta + 2\pi k)\right]$$

where $k = 0, 1, -1, 2, -2, 3, -3, ...$
Figure 10.8 shows the values of $k = 1$ and $k = 2$.

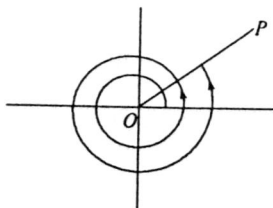

Figure 10.8

Each value of k is just a rotation of the line OP through an angle 2π. So in general $(1+i)$ can be written as

$$1+i = \sqrt{2}\left[\cos\left(\frac{\pi}{4} + 2\pi k\right) + i\sin\left(\frac{\pi}{4} + 2\pi k\right)\right]$$

Take cube roots of each side and apply de Moivre's theorem,

$$(1+i)^{\frac{1}{3}} = 2^{\frac{1}{6}}\left[\cos\left(\frac{\pi}{12} + \frac{2\pi k}{3}\right) + i\sin\left(\frac{\pi}{12} + \frac{2\pi k}{3}\right)\right]$$

Substituting in some values for k and simplifying we obtain

$k = 0$ $\qquad$ $(1+i)^{\frac{1}{3}} = 1.084 + 0.291i$

$k = 1$ $\qquad$ $(1+i)^{\frac{1}{3}} = -0.794 + 0.794i$

$k = 2$ $\qquad$ $(1+i)^{\frac{1}{3}} = -0.291 - 1.084i$

$k = 3$ $\qquad$ $(1+i)^{\frac{1}{3}} = 1.084 + 0.291i$

$k = 4$ $\qquad$ $(1+i)^{\frac{1}{3}} = -0.794 + 0.794i$

$k = 5$ $\qquad$ $(1+i)^{\frac{1}{3}} = -0.291 - 1.084i$

The formula for $(1+i)^{\frac{1}{3}}$ gives exactly three different roots corresponding to $k = 0$, 1, 2. When $k = 3$, 4, 5 or $k = 6$, 7, 8 etc the roots repeat themselves. The three roots are shown in the Argand diagram, Figure 10.9.

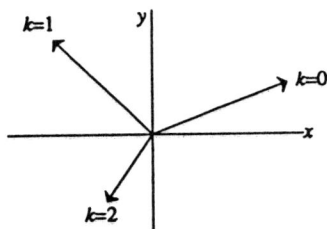

Figure 10.9   The three cube roots of $1 + i$

In general, to find the n n^{th} roots of a complex number z we write z in polar form

$$z = r\left[\cos(\theta + 2\pi k) + i\sin(\theta + 2\pi k)\right]$$

Then

$$z^{\frac{1}{n}} = r^{\frac{1}{n}}\left[\cos\left(\frac{\theta}{n} + \frac{2\pi k}{n}\right) + i\sin\left(\frac{\theta}{n} + \frac{2\pi k}{n}\right)\right]$$

choosing $k = 0$, 1, 2, 3,.. $(n-1)$.

░░░░░░░░░░░░░░░ **Example 10H** ░░░░░░░░░

Find the four fourth roots of $3 + 4i$.

Solution

The modulus of $(3+4i)$ is 5 and the argument of $(3+4i)$ is 0.9273, so that

$$3 + 4i = 5[\cos(0.9273+2\pi k) + i\sin(0.9273+2\pi k)]$$

Take the fourth roots of each side

$$(3+4i)^{\frac{1}{4}} = 5^{\frac{1}{4}}\left[\cos\left(\frac{0.9273}{4}+\frac{2\pi k}{4}\right) + i\sin\left(\frac{0.9273}{4}+\frac{2\pi k}{4}\right)\right]$$

For $k = 0$, 1, 2 and 3 we have the four fourth roots as

$k = 0$	$z_1 = 1.4553 + 0.3436i$
$k = 1$	$z_2 = -0.3436 + 1.4553i$
$k = 2$	$z_3 = -1.4553 - 0.3436i$
$k = 3$	$z_4 = 0.3436 - 1.4553i$

░░░░░░░░░░░░░░░ **Exercise 10E** ░░░░░░░░░

1.   Find the square roots and cube roots of the following complex numbers, giving your answers in cartesian form.

    (a) $1 - i$ (b) $2 + i$ (c) $3i$

    (d) 4 (e) $-1+\sqrt{3}i$ (f) $-5 - 12i$

2.   Find the following complex numbers in polar form.

    (a) $(\sqrt{3}-i)^{\frac{1}{3}}$ (b) $(1+i)^{\frac{4}{3}}$

    (c) $(-3+4i)^{-\frac{2}{3}}$ (d) $(-1)^{\frac{1}{6}}$

3.   If you use the TI-92 to evaluate $(2+i)^{\frac{1}{3}}$ you get the complex number $1.29207 + 0.201294i$. Use an Argand diagram to deduce the other two cube roots.

4.   Use the TI-92 and an Argand diagram to find the following complex numbers.

     (a) $(3i)^{\frac{1}{4}}$ (b) $(\sqrt{3}-i)^{\frac{1}{6}}$

Answers to Exercises

Exercise 1A

2. (a) $y = 3x$ (b) $y = \frac{4}{3}x - \frac{1}{3}$ (c) $y = x + 4$

    (d) $y = -2x + 6$ (e) $y = \frac{1}{3}x - 1$ (f) $y = -\frac{1}{5}x + 5$

    (g) $y = -3x + 7$ (h) $y = -\frac{4}{3}x + \frac{14}{3}$

Exercise 1B

1. (a) No   (b) $S = 10R$ (c) No   (d) $t = 1.5v$
2. $k = \frac{5}{8}$, 6.25N
3. $k = \frac{1}{5}$, 500 miles
4. 1.5 cm

5.
Normal Reaction	$R(N)$	19.8	16.2	15.3
Friction	$F(N)$	6.1	4.99	4.7

    $F = 0.308R$

6. $P = k\rho T$

Exercise 1C

1. (a) $y = 4.03846x + 0.676923$ (b) $s = 3.1r - 4.2$
    (c) $v = -9.8t + 20$ (d) $p = 1.72x + 0.31$

2. (b) $R = 0.172762T + 44.6183$ (c) $R = 48.07$ ohms

3. $v = 3.26743t - 0.048$
    acceleration $= 3.27$ ms^{-2} (to 3 sf)

Exercise 1D

1. (a) $x = 5$ (b) $x = 5.5$ (c) $x = \frac{6}{7}$ (d) $x = \frac{2}{5}$

    (e) $x = 1$ (f) $x = -3$ (g) $x = 2$ (h) $x = \frac{6}{5}$

    (i) $t = 4$ (j) $t = 26$ (k) $x = 1.575$ (l) $x = 1.39273$

2. 45, 47, 49

3. (i) length $= 20$ cm, width $= 10$ cm
    (ii) length $= 19$ cm, width $= 11$ cm

4.   (a) $a = 2$ (b) $u = 4$ (c) $u = -2$

Exercise 1E

1.   (a) $x = 10, y = 6$ (b) $x = 3, y = 2$ (c) $x = 6, y = 2$

      (d) $x = \dfrac{38}{17}, y = -\dfrac{20}{17}$ (e) $a = 4, b = -2$ (f) $u = -2, t = 5$

Exercise 1F

2.   (a) $(x+4)(x+5)$ (b) $(x-4)(x-3)$ (c) $(x+1)(x+3)$ (d) $(x+2)(x+3)$

      (e) $(x-1)(x+1)$ (f) $(2x+1)(x-5)$ (g) $(2x+1)(x+2)$ (h) $(5x+1)(x-7)$

      (i) $(3x+2)(x-4)$ (j) $(x-2)(x+2)$

3.   (a) $x = -3$ (repeated)   (b) $x = -2, x = -8$
      (c) $x = 2, x = -2.5$ (d) $x = 4, x = -4$

4.   (a) $x = 1, x = -7$ (b) $x = \dfrac{-2 - \sqrt{31}}{3}, x = \dfrac{-2 + \sqrt{31}}{3}$

      (c)   No real solutions   (d)   No real solutions

      (e) $x = \dfrac{5 + \sqrt{13}}{6}, x = \dfrac{5 - \sqrt{13}}{6}$ (f)   No real solutions

      (g) $x = 1, x = -\dfrac{94}{13}$ (h) $x = \sqrt{7}, x = -\sqrt{7}$

Exercise 1H

1.   (a)   range $-27 \leq y \leq 125$, one-to-one
      (b)   range $0 \leq y \leq 36$, many-to-one
      (c)   range $0 \leq y < 8$, one-to-one
      (d)   range $-\infty < y \leq -\frac{4}{3}$ and $\frac{4}{5} \leq y < \infty$, one-to-one
      (e)   range $0 \leq y \leq 626$, many-to-one
      (f)   range $0 \leq y \leq 5$, many-to-one

2.   (a) 0   (b) 1   (c) 3   (d) 4   (e) 5   (f) -7 (g) y^2
      (h) $\dfrac{u}{2}$ (i) $y^2 - 6$

3. (a) $\dfrac{x^2}{4}$ (b) $\dfrac{x-6}{2}$ (c) $\dfrac{x^2}{2}$

   (d) $\dfrac{x}{2}-6$ (e) $\left(\dfrac{x-6}{2}\right)^2$ (f) $\dfrac{x^2-6}{2}$

Exercise 1I

1. (a) $f^{-1}(x)=\dfrac{x+10}{6}$ (b) $f^{-1}(x)=\dfrac{x}{4}-2$ (c) $f^{-1}(x)=\dfrac{x^2-7}{2}$

   (d) $f^{-1}(x)=\sqrt[3]{\dfrac{x+5}{8}}$ (e) $f^{-1}(x)=\dfrac{x^2+4}{2}$ (f) $f^{-1}(x)=\left(\dfrac{x+5}{4}\right)^2-1$

   (g) $f^{-1}(x)=2((x+5)^2-1)$ (h) $f^{-1}(x)=\sqrt[3]{(5x)^2-4}$

2. (a) $f^{-1}(x)=\dfrac{2(x+2)}{1-x}$ (b) $f^{-1}(x)=\dfrac{x}{1-x}$ (c) $f^{-1}(x)=\sqrt[3]{\dfrac{2x}{1-x}}$

   (d) $f^{-1}(x)=\dfrac{3}{2x}$ (e) $f^{-1}(x)=\dfrac{2(x+3)}{1-x}$ (f) $f^{-1}(x)=\sqrt[3]{\dfrac{3-x}{2x-1}}$

3. (a) 0   (b) $\frac{1}{2}$ (c) 4.5   (d) $-\frac{47}{8}$ (e) -6.5 (f) $\frac{5}{2}$

Exercise 1J

(a) $x=1, y=0$ (b) $x=-\frac{2}{3}, y=0$ (c) $x=1, x=2, y=0$

(d) $x=0, x=-2, y=0$ (e) $x=-1, y=2$ (f) $x=4, y=1$

(g) $x=-2, x=4, y=0$ (h) $x=5, x=-4, y=1$

Exercise 1K

1. (a) yes   (b) no   (c) yes   (d) no

2. vertical asymptotes $x=-2, x=2$
   horizontal asymptotes $y=4, y=-4$
   domain $-\infty<x<-2, 2<x<\infty$
   range $-\infty<y<-4, 4<y<\infty$

Exercise 2A

1. (a) x^5 (b) a^{11} (c) $128a^5$

    (d) $3x^3$ (e) c^4 (f) $3^8 = 6561$
    (g) a^6 (h) $a^8 b^{12}$ (i) $\dfrac{1}{5}$
    (j) $\dfrac{1}{2^4} = \dfrac{1}{16}$ (k) a^2 (l) $a^{-1} = \dfrac{1}{a}$
    (m) mg^3 (n) $a^{3x} - a^{-x}$ (o) $\dfrac{a^6}{x^4 y^2}$

2. (a) $x^{\frac{1}{2}}$ (b) $x^{\frac{5}{2}}$ (c) $x^{\frac{1}{2}}$
    (d) x^6 (e) x^2 (f) x^{-5}
    (g) x^5 (h) $x^{-\frac{1}{2}}$ (i) $x^{-\frac{1}{2}}$

3. (a) 3   (b) 2   (c) 2
    (d) 1000   (e) 10   (f) $\frac{1}{2}$
    (g) 1   (h) 5   (i) 1

4. (a) 320000   (b) 0.0001473   (c) 9810
    (d) 0.00000103   (e) 192300000000   (f) 3168000
    (g) 59050000000   (h) 4740000

5. (a) 3.75×10^8 joules   (b) 3.16×10^{-11} m
    (c) $2.7 \times 10^{-3} m$ (d) 1.82×10^{-22} joules
    (e) 6.3×10^5 N   (f) 3.15×10^7 s

Exercise 2B

1. (a) 1   (b) 7.389   (c) 0.04979
    (d) 4.953   (e) 1.234   (f) 1.822
    (g) 2.718   (h) 1.649   (i) 54.60
    (j) 0.8187

3. (a) 1.96051   (b) 8.16617
    (c) 1210.29   (d) −0.0216608

5. rate of change of $y = 3y$ $(= 3e^{3x})$

Exercise 2C

1.   (a) 0.30103   (b) 1   (c) 0.6234
      (d) −0.1549   (e) 0.6931   (f) 1.435
      (g) −0.35667   (h) 0   (i) 3
      (j) 2   (k) 0   (l) 1

3.   (a) $x = 0.491$ (b) $x = -0.699$ (c) $x = 0.531$
      (d) $x = 1.629$ (e) $t = 2.416$ (f) $t = -0.755$

      (g) $t = 5.474$ (h) $x = 66.69$ (i) $x = \dfrac{10^6}{3}$

      (j) $t = 4.440$

4. Q_0 is the maximum change
    $t = 0.014$ seconds
    As $t \to \infty$, $Q \to Q_0$

5. $t = 2.77259$

6.   5599 years
    18599 years
    24198 years

7.   (a) 86071, 74082, 47237, 22313
    (c) 4.62 km
    (d) 15.35 km

Exercise 2D

1.   (a) $3\ln a + 2\ln b$ (b) $2\ln x + \ln 0.1$
    (c) $\ln 3.7 + \ln t - \ln a$ (d) $\ln 3 + \ln p + 2\ln(x+y)$
    (e) $\tfrac{1}{2}\log 2 + \log s - \log t$ (f) $\log 0.2 + 2\log v$
    (g) $1 + \log x + 2\log y$ (h) $\ln a + 2\ln t - \ln 2$
    (i) $\log 3 + \log(a+b) - \log s$

2.   (a) $\log 8$ (b) $\log(20x)$

    (c) $\ln\left(\dfrac{10ab}{c}\right)$ (d) $\ln(a^3 b^5)$

    (e) $\ln\left(\dfrac{x^2}{y^4}\right)$ (f) $\log\left(\dfrac{a^{1.5}x}{y}\right)$

3.   (a) $x = 135.2$ (b) $x = 2.008$ (c) $x = 2.686$

Exercise 2E

1. (a) linear law 6
   (b) power law 1
   (c) exponential law 2,3

2. (a) $y = 3.43x^{1.9}$ (b) $t = 1.5s^{-2}$

3. (a) $v = 4e^{-0.2u}$ (b) $y = 6.8e^{1.2x}$

4. $T = 0.2R^{1.5}$

5. S.V.P. $= 0.61e^{0.067T}$

6. $I = 0.00076e^{24.5V}$

7. $p = 100e^{-0.15h}$

8. $k = 5.1 \times 10^{12}e^{-30300/T}$

Exercise 3A

1. (a) $\dfrac{13\pi}{36} = 1.13\,\text{radians}$ (b) $\dfrac{\pi}{10} = 0.314\,\text{radians}$
   (c) $85.9°$ (d) $28.6°$
   (e) $160.4°$ (f) $\dfrac{5\pi}{6} = 2.618\,\text{radians}$
   (g) $286.5°$ (h) $\dfrac{31\pi}{20} = 4.869\,\text{radians}$
   (i) $57.296°$ (j) $\dfrac{\pi}{180} = 0.0175\,\text{radians}$

2.

Degrees	0	30	45	60	90	120	135	150	180
Radians	0	$\frac{\pi}{6}$	$\frac{\pi}{4}$	$\frac{\pi}{3}$	$\frac{\pi}{2}$	$\frac{2\pi}{3}$	$\frac{3\pi}{4}$	$\frac{5\pi}{6}$	π

Degrees	210	225	240	270	300	315	330	360
Radians	$\frac{7\pi}{6}$	$\frac{5\pi}{4}$	$\frac{4\pi}{3}$	$\frac{3\pi}{2}$	$\frac{5\pi}{3}$	$\frac{7\pi}{4}$	$\frac{11\pi}{6}$	2π

3. (a) 0.4 radians   (b) 1.75 radians

4. $\frac{2}{3} = 0.67$ radians; 48 cm^2

5. 1.2 cm

TI-92 ACTIVITY 3A

(E) (a) 10.30 m   (b) 29.1°

(F) 6.71 miles, 26.6°

(G) 58.4 m

(H) (a) 36.9°   (b) 63.4°   (c) 63.5°

TI-92 ACTIVITY 3B

(C) $a = 5.76$ cm, $b = 9.22$ cm   (D) $A = 39.1°$, $C = 64.1°$, area = 32.1 cm^2

(E) (a) 622 m   (b) 65 m

(F) (a) $PR = 125.7$ m, $QR = 178.2$ m   (b) area = 9284 m^2

(G) 170° 34.9 cm   (H) 4.27 nautical miles, bearing 333°

Exercise 3B

2. (a) odd   (b) even   (c) odd   (d) neither
    (e) neither   (f) even   (g) even   (h) odd
    (i) even   (j) even   (k) even

Exercise 3C

2. (a) 57.289962   (b) 572.95721   (c) 5729.5779
    (d) −5729.5779   (e) −572.95721   (f) −57.289962

3. At every odd multiple of 90° ie. ±270°, ±450°, ±630° .....

Exercise 3D

1. (a) (i) 3   (ii) 5   (iii) 0   (iv) 0

    (b) (i) 4   (ii) 1   (iii) $\dfrac{\pi}{2}$ (iv) 0

    (c) (i) 1   (ii) 3   (iii) $-\dfrac{2\pi}{3}$ (iv) 0

    (d) (i) 1   (ii) 3   (iii) 0   (iv) 2
    (e) (i) 1   (ii) 3   (iii) $-\frac{2}{3}$ (iv) 0

    (f) (i) 2   (ii) 4   (iii) $\dfrac{\pi}{4}$ (iv) 1

    (g) (i) 1   (ii) $\frac{1}{2}$ (iii) 2π (iv) -3

    (h) (i) 0.1   (ii) 2π (iii) $\frac{3}{2}$ (iv) 0.5

3. (a) $6\cos 2t$ (b) $3\sin(4t+\pi)$ (c) $\tan 5t + 3$

    (d) $10\sin\left(5t - \dfrac{\pi}{2}\right)$ (e) $0.5\cos(3t - 2) - 1$

    (f) $\dfrac{\pi}{10}$ (g) $\dfrac{10}{\pi}$ (h) 200π

5. (a) (i) $\dfrac{2\pi}{3}$ (ii) $\dfrac{3}{2\pi}$

    (b) (i) $\dfrac{2\pi}{5}$ (ii) $\dfrac{5}{2\pi}$

    (c) (i) $\dfrac{1}{4}$ (ii) 4

    (d) (i) 4   (ii) $\dfrac{1}{4}$

    (e) (i) $\dfrac{\pi}{4}$ (ii) $\dfrac{4}{\pi}$

7. (a) $\lim\limits_{x\to 0}\dfrac{x}{\sin x} = 1$ (b) $\lim\limits_{x\to 0}\dfrac{\sin x}{x^2}$ is undefined

    (c) $\lim\limits_{x\to 0}\dfrac{1 - \cos x}{x} = 0$ (d) $\lim\limits_{x\to 0}\dfrac{\sin x - x\cos x}{x^3} = \dfrac{1}{3}$

Exercise 3E

1. $f(t) = 80\sin 120\pi t$

2. 6 cm, 1.5 Hz, at the central position, 1.85 cm to the left of the central position

3. $T = 11\sin\left(\dfrac{\pi t}{6} + \dfrac{\pi}{2}\right) + 16$, in June $T = 21.5,°C$, in January $T = 6.47°C$

4. $a = 6.5$, $w = \dfrac{2\pi}{12.4} = 0.5067$, $\alpha = \dfrac{\pi}{2}$
      h = 5.97 m, 4.48 m, 2.26 m

Exercise 3F

1.   23.6°   2.   66.4°   3.   21.8°
4.   −53.1°   5.   143.1°   6.   −38.7°

Exercise 3G

1.   (a) $x = -331.965°, -208.035°, 28.034°, 151.965°$
      (b) $x = -260.212°, -99.788°, 99.788°, 260.212°$
      (c) $x = -305.54°, -125.54°, 54.46°, 234.46°$
      (d) $x = -312.84°, -47.16°, 47.16°, 312.84°$
      (e) $x = -243.43°, -63.43°, 116.57°, 296.57°$
      (f) $x = -117.13°, -62.87°, 242.87°, 297.13°$

3.   (a) $\theta = -71.94°, 71.94°, 288.06°, 431.94°$
      (b) $\theta = -41.67°, 138.33°, 318.33°, 498.33°$
      (c) $\theta = 64.16°, 115.84°, 424.16°, 475.84°$

5.   (a) $t = -5.20168, -1.0815, 1.0815, 5.20168$
      (b) $t = -6.11946, -3.30532, 0.16373, 2.97786$
      (c) $t = -3.23135, -0.0897581, 3.05183, 6.19343$
      (d) $t = -3.00113, -0.140461, 3.28205, 6.13858$
      (e) $t = -5.36809, -2.22649, 0.9151, 4.05669$
      (f) $t = -4.72238, -1.56079, 1.56079, 4.72238$

Exercise 3H

1.   (a)   51.7 °   (b)   −46.4 °   (c)   −61.6 °
      (d)   81.4 °   (e)   3.44 °   (f)   161.8 °

2.   (a)   −1.29   (b)   −0.100   (c)   1.51
      (d)   1.91   (e)   0.464   (f)   0.110

3.   (a)   0.995   (b)   0.995   (c)   0.75
      (d)   −0.314   (e)   0.298   (f)   −0.848
      (g)   0.8   (h)   −0.55   (i)   0.75

Exercise 3I

3. (a) $x = 45°, 135°, 215°, 315°$
    (b) $\theta = 15.3°, 164.7°$
    (c) $x = 35.9°, 144.1°, 226.9°, 313.1°$
    (d) $x = 60°, 131.8°, 228.2°, 300°$
    (e) $x = 53.6°, 147.5°, 212.5°, 306.4°$
    (f) $x = 61°, 119°, 270°$
    (g) no solutions
    (h) $\theta = 39.2°, 140.8°, 219.2°, 320.8°$

Exercise 3J

1. $\cos(A+B) = 0$
    $\sin(A+B) = 1$
    $\tan(A+B)$ is undefined
    $A + B = 90°$

2. (a) $\dfrac{7}{25}$ (b) $\dfrac{15}{17}$ (c) $\dfrac{7}{24}$ (d) $\dfrac{15}{8}$

    (e) $-\dfrac{304}{425}$ (f) $\dfrac{304}{297}$ (g) $\dfrac{87}{425}$ (h) $\dfrac{416}{87}$

    (i) $\dfrac{527}{625}$ (j) $\dfrac{240}{289}$

3. (a) 0.866 (b) 0.6 (c) 1.73 (d) 0.75
    (e) 0.393 (f) −0.427 (g) −0.1196 (h) −8.30
    (i) −0.5 (j) 0.96

4. (b) (i) $-\sin A$ (ii) $-\cos A$ (iii) $-\sin A$
        (iv) $-\cos A$ (v) $\sin A$ (vi) $\dfrac{\tan A - 1}{1 + \tan A}$

5. (a) $-\sin 7x$ (b) $2\sin 40°\cos 20°$ (c) $-2\sin 40°\sin 10°$
    (d) $\frac{1}{2}(\sin 80° + \sin 20°)$ (e) $-\frac{1}{2}(\cos 50° - \cos 30°)$ (f) $\sin 29°$
    (g) $2\cos 40°\sin 5°$ (h) $\sin(45°+x)$

6. (b) $4\cos^3 A - 3\cos A$ (c) $4\sin A \cos A - 8\sin A \cos^3 A$

Exercise 3K

1.   (a) $13\sin(x+22.6°)$ (b) $13\sin(x–22.6°)$
      (c) $\sqrt{5}\sin(x+63.4°)$ (d) $10\sin(x–53.1°)$
      (e) $\sqrt{26}\sin(x-11.3°)$ (f) $\sqrt{45}\sin(x-26.6°)$

2. $A\sin x + B\cos x = R\cos(x-\alpha)$ where $R = \sqrt{A^2 + B^2}$ and $\alpha = \tan^{-1}\left(\dfrac{A}{B}\right)$

3.   (a) $13\cos(-22.6°)$ (b) $\sqrt{45}\cos(x-116.6°)$
      (c) $\sqrt{5}\cos(x-116.6°)$ (d) $\sqrt{10}\cos(x-18.4°)$

4.   (a) $x = 4.9°, 129.9°$ (b) $x = -157.4°, 22.6°$
      (c) $x = -36.8°, 90°$ (d) $x = 143.1°$
      (e) $x = -180°, 22.6°, 180°$ (f) $x = 74.8°, 158.4°$

Exercise 4A

1.   (i) $u_n = 3n + 1$ (ii) $u_n = n^2 + 2$
      (iii) $u_n = n^2 - 1$ (iv) $u_n = 5n - 1$

2.   (a)   2, 4, 8, 16, 32   no limit
      (b) $1, \dfrac{1}{3}, \dfrac{1}{9}, \dfrac{1}{27}, \dfrac{1}{81}$ limit = 0
      (c) $-2, 4, -8, 16, -32$ no limit
      (d) $-\dfrac{1}{2}, \dfrac{1}{4}, -\dfrac{1}{8}, \dfrac{1}{16}, -\dfrac{1}{32}$ limit = 0
      (e) $3, \dfrac{5}{2}, \dfrac{7}{3}, \dfrac{9}{4}, \dfrac{11}{5}$ limit = 2
      (f)   1.1, 2.01, 3.001, 4.0001, 5.00001   no limit

3.   (a) $u_n = 3^{n-1}$ (b) $u_n = 19 - 5n$
      (c) $u_n = 5 + (0.1)^n$ limit = 5   (d) $u_n = \dfrac{n}{n+2}$ limit = 1

Exercise 4B

1.   (a) $4 + 9 + 16 + 25 = 54$ (b) $1+\dfrac{1}{2}+\dfrac{1}{3}+\dfrac{1}{4}+\dfrac{1}{5}+\dfrac{1}{6}=\dfrac{49}{20}=2.45$
      (c) $2 + 6 + 10 + 14 = 32$ (d) $2 + 4 + 6 + 8 + 10 = 30$
      (e) $1+\dfrac{1}{4}+\dfrac{1}{9}=\dfrac{49}{36}$ (f) $-2 + 5 + 24 + 61 = 88$

2.   (a) $\sum_{i=1}^{5}(21-5n)$ (b) $\sum_{i=1}^{4}2(3^{i-1})$

      (c) $\sum_{i=1}^{6}\frac{1}{2}i^2$ (d) $\sum_{i=1}^{5}(-2)^{i-1}$

3.   only (e) to $\dfrac{\pi^2}{6}$

Exercise 4C

1.   (a) 64   (b) -3 (c) 19
           $u_n = 9n - 8$ $u_n = 5 - n$ $u_n = 2n + 3$

2.   (a) $d = 3, n = 28$ (b) $d = 4, n = 18$ (c) $d = -7, n = 16$

3.   (a) 852   (b) -78 (c) 1515

4.   (a) 100   (b) 205

5.   420

6.   4

7. $a = 6, d = 4$

Exercise 4D

1.   (a) (i)   1.2   (ii) 20.6391   (iii) 65.9963
      (b) (i)   0.8   (ii) 0.174483   (iii) 5.40948
      (c) (i) -0.5 (ii) 0.0033203   (iii) -1.1289
      (d) (i)   1.6   (ii) 13743.9   (iii) 13983.3

2.   (a) $66\frac{2}{3}$ (b) does not exist   (c) 10   (d) $-\dfrac{10}{19}$

3. $2\frac{1}{3}$ m

4.   2 or $\dfrac{1}{3}$

Exercise 4E

1. (a) $x^4 + 4x^3 + 6x^2 + 4x + 1$ (b) $8 + 12x + 6x^2 + x^3$

(c) $1 - 5x + 10x^2 - 10x^3 + 5x^4 - x^5$ (d) $625 + 1500x + 1350x^2 + 540x^3 + 81x^4$

(e) $125 - 150x + 60x^2 - 8x^3$ (f) $x^5 - 5x^4y + 10x^3y^2 - 10x^2y^3 + 5xy^4 - y^5$

(g) $u^4 + 4u^2 + 6 + \dfrac{4}{u^2} + \dfrac{1}{u^4}$ (h) $16x^4 + 32x^2 + 24 + \dfrac{8}{x^2} + \dfrac{1}{x^4}$

2. (a) 1920000 (b) −96 (c) −343 (d) −12

3. (a) 1716 (b) −42240 (c) 9773.16 (d) -6.08402×10^{-6}
(e) $6435a^8b^7$ (f) $-3432a^7b^7$

4. (a) $(1+x)^6$ (b) $(0.5+x)^4$ (c) $(1-0.5x)^4$

Exercise 4F

1. (a) $1 - 2x + 3x^2 - 4x^3$ (b) $1 - 3x + 6x^2 - 10x^3$

(c) $1 + \dfrac{3}{2}x + \dfrac{3}{8}x^2 - \dfrac{1}{16}x^3$ (d) $1 - \dfrac{x}{2} + \dfrac{3x^2}{8} - \dfrac{5}{16}x^3$

2. (a) $|x| < \dfrac{1}{2}$ (b) $|x| < 3$ (c) $|x| < \dfrac{1}{4}$ (d) $|x| < \dfrac{1}{5}$

3. (a) $1 + \dfrac{3}{2}x - \dfrac{9}{8}x^2 + \dfrac{27}{16}x^3$ $|x| < \dfrac{1}{3}$

(b) $1 + 4x + 12x^2 + 32x^3$ $|x| < \dfrac{1}{2}$

(c) $1 - 4x + 16x^2 - 64x^3$ $|x| < \dfrac{1}{4}$

(d) $1 - x + \dfrac{3}{2}x^2 - \dfrac{5}{2}x^3$ $|x| < \dfrac{1}{2}$

4. 0.989949, 1.0955

5. (b) $1 - \dfrac{6}{x} + \dfrac{27}{x^2} - \dfrac{108}{x^3}$ (c) $|x| < 3$ (d) $\dfrac{1}{x^2} - \dfrac{6}{x^3} + \dfrac{27}{x^4} - \dfrac{108}{x^5}$

6.   (a) $\dfrac{1}{2} - \dfrac{x}{4} + \dfrac{x^2}{8} - \dfrac{x^3}{16}$ (b) $\dfrac{1}{4} - \dfrac{x}{8} + \dfrac{x^2}{16} - \dfrac{x^3}{32}$

     (c) $\dfrac{1}{25} - \dfrac{2}{125}x^2 + \dfrac{3}{625}x^2 - \dfrac{4}{3125}x^3$ (d) $\sqrt{2}\left(1 + \dfrac{1}{4}x - \dfrac{1}{32}x^2 + \dfrac{1}{128}x^3\right)$

Exercise 5A

1. ± 1.50 2. -2.31 3. 1.935 4. -0.567

5.   (a) 2.15443 (b) 2.71442

Exercise 5B

1. 6.1926, no

2. 0.8074, no

3.   (b) 3.1004, 3.107
     (c)   2nd converges faster, 3.1073

4.   (a) 2.8284, 3.1623
     (c) $2x^2 = a$, 2.2361

5.   (a) 0
     (b) $x_{n+1} = \sqrt{\sin x_n}$, $x_{n+1} = \sin^{-1}(x_n^2)$, 1st converges
     (c) 0.877

6.   (a) 1.74
     (b) 0.739
     (c) -1.841, 1.146

Exercise 6A

1.   (a) $7x^6$ (b) $\dfrac{1}{3}x^{-\frac{2}{3}}$ (c) $-3x^{-4}$ (d) $-\dfrac{5}{2}x^{-\frac{7}{2}}$

     (e) $12x^2$ (f) $9x^{\frac{1}{2}}$ (g) $-\dfrac{0.7}{x^2}$ (h) $99x^8$

     (i) 0 (j) $30x^4$ (k) $-3x^{-2}$ (l) 0

2.   (a) 12 (b) 5 (c) 1
     (d) -4 (e) $-\dfrac{1}{9}$ (f) $\dfrac{1}{4}$

Exercise 6B

1. (a) $30x^4 - 2$ (b) $-3x^{-2} - 4x^{-3}$ (c) $15 - 12x^3 + 14x^6$
   (d) $-0.3x^{-4} + 5.7x^2$ (e) $0.8x^{-0.8} + 0.5x^{-1.1}$ (f) $-0.3x^{-4} - 3.8x^{-3}$

   (g) $1.8x - 0.4x^3$ (h) $-4.2x^{-2} + 1.05x^{-\frac{1}{2}}$ (i) $\dfrac{1}{3}x^{-\frac{2}{3}} + \dfrac{1}{2}x^{-\frac{1}{2}}$

2. (a) $6, y = 6x - 9$ (b) $12, y = 12x + 16$

   (c) $\dfrac{5}{16}, y = \dfrac{5}{16}x - \dfrac{1}{8}$ (d) $\dfrac{1}{4}, y = \dfrac{1}{4}x + 1$

   (e) $\dfrac{5}{4}, y = \dfrac{5}{4}x - \dfrac{1}{4}$ (f) $0.461827, y = 0.461827x + 0.791704$

   (g) $-\dfrac{100}{9}, y = -\dfrac{100}{9}x + \dfrac{20}{3}$ (h) $-1.11976, y = -1.11976x + 2.05094$

   (i) $-96, y = -96x - 144$ (j) $135, y = 135x - 729$
   (k) $-2, y = -2x + 1$ (l) $-7, y = 10 - 7x$

3. (a) -1, decreasing; 15, increasing
   (b) 12, increasing; 0, stationary; 3 increasing
   (c) -7, decreasing; -3.25, decreasing
   (d) -1.21837, decreasing; 1.11679, increasing
   the curve reaches a local minimum

Exercise 6C

1. 40mph
   Getting to and from the M1
   Road works on the M1
   Stops at service stations

2. 1.15ms^{-2}

3. (a) 35ms^{-1} (b) 25ms^{-1} (c) 17.5ms^{-1}

   (b) $40\text{ms}^{-1}, 30\text{ms}^{-1}, 20\text{ms}^{-1}, 10\text{ms}^{-1}, 0\text{ms}^{-1}$
   The stone stops moving instantaneously at its highest point

   (c) -10ms^{-2} for all times; the acceleration is directed downwards

4. 1.41 fish per second; 0.67 fish per second

5. $\dfrac{dA}{dr} = 2\pi r$

6. (a)

x	$f(x)$	$\Delta f(x)$	av. rate
1.7	7.87	1.12	5.6
1.6	7.28	0.53	5.3
1.55	7.0075	0.2575	5.15
1.51	6.80030	0.0502999	5.02999
1.501	6.75500	0.05	5
1.5001	6.7505	0.0005	5

Instantaneous rate of change of f when $x = 1.5$ is 5

(b) $\dfrac{df}{dx} = 5$

7.

r	$p(r)$	$\Delta p(r)$	av. rate
2.2	4.78181	0.78181	3.90905
2.1	4.39523	0.39523	3.95229
2.01	4.03995	0.03995	3.995
2.001	4.00399	0.00399	3.99
2.0001	4.0004	0.0004	4

The instantaneous rate of change of p when $r = 2$ is 4

Exercise 6D

1. (a) $2x$ (b) 1 (c) $6x - 4$ (d) $10x + 2$

   (e) $3x^2$ (f) $6x^2 + 3$ (g) $-\dfrac{1}{x^2}$ (h) 0

2. $g'(t) = \lim\limits_{h \to 0} \dfrac{g(t+h) - g(t)}{h}$

   (a) 1 (b) $6t - 4$ (c) $15t^2 - 6$ (d) 0

3. (a) $\lim\limits_{h \to 0} \dfrac{\{(u+h)^2 - 3(u+h) + 5\} - \{u^2 - 3u + 5\}}{h} = 2u - 3$

   (b) $\lim\limits_{h \to 0} \dfrac{\{(s+h)^3 + (s+h)\} - \{s^3 + s\}}{h} = 3s^2 + 1$

   (c) $\lim\limits_{h \to 0} \dfrac{\{-0.5(t+h)^2 + 10(t+h) - 1\} - \{-0.5t^2 + 10t - 1\}}{h} = -t + 10$

(d) $\lim\limits_{h\to 0}\dfrac{\left\{\dfrac{1}{(u+h)}+4(u+h)\right\}-\left\{\dfrac{1}{u}+4\right\}}{h}=-\dfrac{1}{u^2}+4$

(e) $\lim\limits_{h\to 0}\dfrac{(x+h)^5-x^5}{h}=5x^4$

Exercise 6E

1. (a) $(0, -12)$ minimum
   (b) $(0, 9)$ maximum
   (c) $(-1.25, -6.125)$ minimum
   (d) $(-0.167, 2.083)$ maximum
   (e) $(0, 3)$ point of inflexion
   (f) $(-0.0972, 5.049)$ maximum $(3.431, -16.9)$ minimum
   (g) $(-0.76929, 10.8765)$ maximum
   (h) $(-1, -4)$ maximum $(1, 0)$ minimum

2. (e) $(0, 3)$
   (f) $(1.67, -5.926)$
   (g) $(0.1835, 2.7542)$ $(1.8165, -10.3097)$

Exercise 6F

1. (a) $6x$
   (b) $-12x + 100x^3 - 30x^4$
   (c) $2+\dfrac{1}{4}x^{-\frac{3}{2}}+\dfrac{5}{16}x^{-\frac{9}{4}}$

   (d) $\dfrac{2}{x^3}$
   (e) $-\dfrac{1}{2}x^{-\frac{3}{2}}+6x$
   (f) $8-210x^4$

   (g) 8
   (h) 0
   (i) $-\dfrac{1}{4}t^{-\frac{3}{2}}+\dfrac{3}{8}t^{-\frac{7}{4}}+\dfrac{4}{3}t^{-\frac{7}{3}}$
   (j) $n(n-1)x^{n-2}$

3. (a) $20x^3 - 24x^2;\ 60x^2 - 48x;\ 120x - 48$

   (b) $-12x + 100x^3 - 30x^4;\ -12 + 300x^2 - 120x^3;\ 600x - 360x^2$

   (c) $\dfrac{2}{x^3};\ -\dfrac{6}{x^4};\ \dfrac{24}{x^5}$
   (d) $-\dfrac{1}{4}t^{-\frac{3}{2}}-12t;\ -12+\dfrac{3}{8}t^{-\frac{5}{2}};\ -\dfrac{15}{16}t^{-\frac{7}{2}}$

Exercise 6G

(a) $(0, -11)$ minimum
(b) $(1, 7)$ inflexion
(c) $(0.382, -1)$ min; $(1.5, 0.5625)$ max; $(2.618, -1)$ min
(d) $(1.42264, 0.385)$ max; $(2.57735, -0.385)$ min
(e) $(1, 2)$ max, $(-1, -2)$ min
(f) none

Exercise 6H

1. 1m from one end

2. Base radius = 1.3365 m
    Height = 2.673 m
    Area required = 33.7 m

3. If $S = x + y$ (and $xy = k$), $\dfrac{dS}{dx}$ is a minimum when $x = y = \sqrt{k}$
    For $k = 10^{-4}$, $x = y = 10^{-2}$ Kmol^{2}dm^{-6}

4. (i) $v = \dfrac{1}{3}\sqrt{\dfrac{T}{3a}}$

5. 800 m^2

7. 0.08

8. $r = \dfrac{20}{3}$ and $V = \pi\left(\dfrac{20}{3}\right)^3$

Exercise 6I

1. (a) $4e^{4x}$, $16e^{4x}$ (b) $-7e^{-7x}$, $49e^{-7x}$

    (c) $2e^{0.5x}$, $e^{0.5x}$ (d) $-2.6e^{-1.3x}$, $3.38e^{-1.3x}$

    (e) $\dfrac{1}{x}$, $-\dfrac{1}{x^2}$ (f) $\dfrac{3}{x}$, $-\dfrac{3}{x^2}$

    (g) $\pi\cos(\pi x)$, $-\pi^2\sin(\pi x)$ (h) $2\cos(2x)$, $-4\sin(2x)$

    (i) $13.02\cos(3.1x)$, $-40.36\sin(3.1x)$ (j) $-4\sin 4x$, $-16\cos 4x$

    (k) $-0.2\sin(0.2x)$, $-0.04\cos(0.2x)$ (l) $-3\pi\sin(2\pi x)$, $-6\pi^2\cos(2\pi x)$

    (m) $0.03e^{0.1x} - 0.35\cos(0.5x)$, $0.03e^{0.1x} + 0.175\sin(0.5x)$

    (n) $-12\sin 3x - 12\cos 4x$, $-36\cos 3x + 48\sin 4x$

    (o) $-0.1e^{-0.1x} + 0.1e^{0.1x}$, $0.01e^{-0.1x} + 0.01e^{0.1x}$

    (p) $\dfrac{1}{x} - \dfrac{6}{x} = -\dfrac{5}{x}$, $\dfrac{5}{x^2}$

2. (a) $y = -5.98x + 3.134$ (b) $y = x - 1$
    (c) $y = 7.389x - 7.389$ (d) $y = -0.1x + 0.4$

4. (a) $-500\,°\mathrm{Cs}^{-1}$ (b) $-3.37\,°\mathrm{Cs}^{-1}$

5. $\dfrac{dP}{dt} = -9.03\,\mathrm{e}^{-2.1t}$

6. (a) $0.21\mathrm{ms}^{-1}$; $0.16\mathrm{ms}^{-1}$
    (b) $0\mathrm{ms}^{-2}$; $-0.0947\mathrm{ms}^{-2}$

Exercise 6J

1. (a) $\dfrac{1}{2\sqrt{x}}\mathrm{e}^{2x} + 2\sqrt{x}\,\mathrm{e}^{2x}$
    (b) $2x\mathrm{e}^{3x} + 3x^2\mathrm{e}^{3x}$

    (c) $5x^4\mathrm{e}^{-2x} - 2x^5\mathrm{e}^{-2x}$
    (d) $\sin 2x + 2x\cos(2x)$

    (e) $\dfrac{\cos(\pi x)}{2\sqrt{x}} - \pi\sqrt{x}\sin(\pi x)$
    (f) $15x^2\mathrm{e}^{3x} + 15x^3\mathrm{e}^{3x}$

    (g) $\sec^2 x$
    (h) $1 + \ln x$

    (i) $\dfrac{2x\cos 2x - 2\sin 2x}{x^3}$
    (j) $\dfrac{\mathrm{e}^{-3x}(1-6x)}{2\sqrt{x}}$

    (k) $\dfrac{-3\sin 2x\sin 3x - 2\cos 3x\cos 2x}{\sin^2 2x}$
    (l) $\dfrac{2\mathrm{e}^{2x}(x-1) - 2\mathrm{e}^{-2x}(x+1)}{x^3}$

    (m) $15(3x-1)^4$
    (n) $2(4x+1)^{-\frac{1}{2}}$

    (o) $-\pi\sin(\pi x - 3)$
    (p) $2x\mathrm{e}^{x^2}$

    (q) $\dfrac{2x}{x^2+1}$
    (r) $-2\tan 2x$

    (s) 0
    (t) $\mathrm{e}^{-2x}(3x^2\cos 3x + 2x(1-x)\sin 3x)$

    (u) $12\sin(1-4x)$
    (v) $\dfrac{3(6x^2-1)^2(6x^2+1)}{x^4}$

    (w) $\mathrm{e}^{5x}(0.7\cos 0.7x + 5\sin 0.7x)$
    (x) $\dfrac{2}{x}$

    (y) $\sec x\tan x$
    (z) $-2(4x+1)^{-\frac{3}{2}}$

2. (a) (0.7937, 1.88988) minimum (b) none
    (c) (0,0) minimum (d) none
    (e) $t = n\pi + 0.294$, $n = 0, \pm 1, \pm 2, \dots$ maximum
         $t = n\pi + 1.865$, $n = 0, \pm 1, \pm 2, \dots$ minimum
    (f) none (g) none (h) inflexions at $n\pi$

3. (c) 1.15 radians = 65.9°

4. (a) 9 cm (b) 1 s (c) 1 cms^{-1} increasing (d) 12 cm

5. (a) (-2, 3) max; (2, 0.33) min

    (b) $x = 2n\pi - \dfrac{\pi}{4}$ min

                                $y = \pm 2.33$

         $x = 2n\pi + \dfrac{3\pi}{4}$ max

    (c) $x = 2n\pi \pm \pi$ points of inflexion

    (d) $x = n\pi + \dfrac{\pi}{4}$ max when n is even
                       min when n is odd

6. (a) $X = R$

Exercise 7A

1. Estimates obtained using interval mid-points. True values in brackets.

    (a) 8.9775 (9)
    (b) 3.74625 (4)
    (c) 0.50028 (0.5)
    (d) 3.59660 (3.69328)

2. (b) distance = 2.37 m (using 10 sub-intervals)

3. (a) 12.92 miles
    (b) measure the speed at shorter time intervals

Exercise 7B

1. (a) $\dfrac{3^5}{5} = \dfrac{243}{5} = 48.6$ (b) $\dfrac{1}{6}$ (c) $\dfrac{1.2^4}{4} = 0.5184$

    (d) $\dfrac{3^5}{5} - \dfrac{1}{5} = \dfrac{242}{5} = 48.4$ (e) $\dfrac{4^6}{2} - \dfrac{2^6}{2} = 2016$

    (f) $\dfrac{4 \times 2^3}{3} - \dfrac{4 \times (-1)^3}{3} = 12$ (g) $-\dfrac{5}{6}$ (h) −3.675

    (i) $\dfrac{144}{77} = 1.8701$ (j) 13.4167

2. (a) $\displaystyle\int_1^4 2x^2\,dx = 42$ (b) $\displaystyle\int_0^1 x - x^4\,dx = 0.3$

    (c) $\displaystyle\int_{-1}^1 2 - x^2\,dx = \dfrac{10}{3}$

3. (b) 2.4 metres

4. (b) $\int_1^4 (20 - x^2)\,dx = 39$

Exercise 7C

1. (a) $e^1 - e^0 = e^1 - 1 = 1.71828$ (b) $\dfrac{e^6}{3} - \dfrac{e^3}{3} = 127.7811$

(c) $\dfrac{e^7}{7} - \dfrac{e^{-7}}{7} = 156.6618$ (d) $\dfrac{e^4}{2} + \dfrac{13}{6} = 29.4657$

(e) $-\dfrac{3e^8}{4} + \dfrac{3e^4}{4} + \dfrac{31}{5} = -2188.57$ (f) 4.7008

(g) $-\dfrac{e^4}{2} + e^2 + \dfrac{3}{2} = -18.41$

2. (a) $\int_0^2 e^x\,dx = e^2 - e^0 = e^2 - 1 = 6.3891$

(b) $\int_{-1}^3 e^{2x}\,dx = \dfrac{e^6}{2} - \dfrac{e^{-2}}{2} = 201.6467$

(c) $\int_0^{1.5} e^{3x} - 1\,dx = 28.1724$

Exercise 7D

1. (a) $x^4 + c$ (b) $\dfrac{1}{2}x^6 + c$ (c) $\dfrac{2}{3}x^{\frac{3}{2}} + c$

(d) $\dfrac{13}{3}x^3 - \dfrac{7}{4}x^4 + c$ (e) $\dfrac{1}{2}x^6 + \dfrac{1}{2}x^4 - \dfrac{1}{2}x^2 + 4x + c$

(f) $6x + \dfrac{3}{2}x^2 - \dfrac{2}{3}x^3 + c$ (g) $\dfrac{16}{3}x^3 + 8x^2 + 4x + c$

(h) $\dfrac{x^3}{3} - x^2 + x + c$ (i) $-\dfrac{1}{x} + c$ (j) $\dfrac{1}{0.3}x^{0.3} + c$

(k) $-\dfrac{1.7}{1.3}x^{-1.3} + c$ (l) $3\ln(x) + c$ (m) $-x^{-5} - 3x^{-1} + c$

(n) $\dfrac{2}{1.3}x^{1.3} + \ln(x) + c$ (o) $\dfrac{1}{2}x^{18} - \dfrac{2}{5}x^5 - 3x^{-1} + c$

2.   (a) $\dfrac{1}{2}e^{2x}+c$ (b) $-\dfrac{1}{5}e^{-5x}+c$ (c) $10e^{0.1x}+c$

    (d) $\dfrac{3}{4}e^{4x}+c$ (e) $e^{6x}+c$ (f) $1.8e^{-0.5x}+c$

    (g) $\dfrac{4}{3}e^{3x}+\dfrac{3}{2}e^{-2x}+c$ (h) $\dfrac{0.6}{3.1}e^{3.1x}+3e^{-0.3x}+c$

3.   (a) $-\dfrac{1}{5}\cos(5x)+c$ (b) $\dfrac{2}{3}\sin(1.5x)+c$

    (c) $-\dfrac{4}{3}\cos(3x)+c$ (d) $-\dfrac{3}{2}\cos(2x)-\dfrac{2}{3}\sin(3x)+c$

    (e) $-\dfrac{2}{\pi}\cos(\pi x)+c$ (f) $\dfrac{0.5}{\pi}\sin(3\pi x)+c$

    (g) $-\dfrac{2}{w}\cos(wx)+c$ (h) $\dfrac{1.5}{7}\sin(7x)-0.15\cos(2x)+c$

    (i) $10e^{0.1x}-\dfrac{2}{\pi}\cos(\pi x)+c$ (j) $x^3-7e^{-0.6x}+\dfrac{1}{0.9}\sin(0.9x)+c$

    (k) $\dfrac{1}{3}\ln(x)-0.1\cos(5x)+c$

4.   (a) $\dfrac{7}{3}(2.33)$ (b) 7.5   (c) $\dfrac{26}{3}(8.67)$

    (d) $\dfrac{2}{3}e^3-\dfrac{2}{3}(12.7236)$ (e) 17.301   (f) $25(e^{0.2x}-e^{-0.2x})\,(10.0667)$

    (g) 2   (h) 1   (i) $\dfrac{178}{3}-70e^{-0.4}\,(12.4109)$

5.   (a) $\dfrac{13}{15}(0.8667)$ (b) 2

    (c) 0.0664   (d) $2e^2-4\,(10.7781)$

6.   (a) $\dfrac{1}{3}x^3-\dfrac{3}{2}x^2+2x+c$ (b) $\dfrac{1}{4}x^4+\dfrac{2}{3}x^3+c$

    (c) $-\dfrac{1}{2}x^{-2}-x^{-1}+c$ (d) $\dfrac{1}{3}x^3-x^{-1}+c$

    (e) $-ax^{-1}+bx+c$ (f) $\dfrac{1}{3}ax^3+\dfrac{1}{2}bx^2+cx+d$

7. $y=2x+x^2-\dfrac{1}{3}x^3-\dfrac{5}{3}$

8. $y=x+\dfrac{1}{3}x^3+1$

9. (a) $\frac{1}{3}t^3 + c$
   (b) $\frac{3}{2}t^2 + t + c$
   (c) $\frac{1}{4}t^4 + \frac{2}{3}t^3 + c$

   (d) $-\frac{1}{t} + c$
   (e) $\frac{5}{8}t^8 + t^4 - \ln(t) + c$
   (f) $\frac{t^3}{3} - t^2 + t + c$

   (g) $\frac{1}{3}at^3 + \frac{1}{2}bt^2 + ct + d$
   (h) $\frac{1}{2}e^{2t} + c$
   (i) $20e^{0.1t} + c$

   (j) $-\frac{2}{\pi}\cos(\pi t) + c$
   (k) $\ln(t) + c$
   (l) $50\sin(0.1t) + c$

   (m) $-\cos(t) + \sin(t) + c$
   (n) $-\frac{1}{v} + c$
   (o) $2p^{\frac{1}{2}} + c$

   (p) $\frac{1}{3}u^3 + \frac{3}{2}u^2 + 8u + c$
   (q) $\ln(w) + c$
   (r) $\frac{1}{2}e^{2u} + c$

   (s) $\frac{2}{3}y^{\frac{3}{2}} + c$

10. (a) $v = \frac{1}{3}t^3 + 2$
    (b) $v = \frac{1}{2}t^2 + t - \frac{1}{2}$
    (c) $v = 3\sin(t) + c$

11. (a) $V = mgx + c$
    (b) $V = \frac{k}{2}x^2 - klx + c$

Exercise 7E

2. (a) $\frac{16}{3}(5.33)$
   (b) $\frac{11}{12}(0.9167)$
   (c) 2
   (d) 39.33

3. (a) -4
   area between $y = x - 2$, $y = 0$, $x = -1$ and $x = 1$ is 4

   (b) 1.3863 (2ln(2))
   area between $y = \frac{1}{x}$, $y = 0$, $x = 0.5$ and $x = 2$ is 1.3863

   (c) 0.6667 $\left(\frac{2}{3}\right)$
   not an area because $y = x^2 - x$ is negative between $x = 0$ and $x = 1$ and positive between $x = 1$ and $x = 2$

   (d) $2(e^1 - e^{-1})$ (4.7008)
   area between $y = 2e^{-x}$, $y = 0$, $x = -1$ and $x = 1$ is 4.7008

(e) 0

not an area because $y = \sin(x)$ is negative between $x = -\dfrac{\pi}{2}$ and $x = 0$ and

positive between $x = 0$ and $x = \dfrac{\pi}{2}$

(f) $1 - 3e^{-1}$ (−0.1036)

not an area because $y = 3e^{-x} - 2$ is negative between $x = \ln(1.5)$ and $x = 1$

4. (a) area = 15.0

$$\int_{-3}^{3} f(x)dx = 15.0$$

(b) area = 15.0

$$\int_{-3}^{3} f(x)dx = 0$$

(c) area = 5.4

$$\int_{0}^{1} f(x)dx = 2.7$$

$$\int_{1}^{2} f(x)dx = -2.7$$

$$\int_{0}^{2} f(x)dx = 0$$

Exercise 7F

1. (a) $x^4 - \dfrac{1}{2}x^2 + c$ (b) $\dfrac{2}{3}(x-1)^{\frac{3}{2}} + c$

(c) $-\dfrac{1}{2}\cos(2x+1) + c$ (d) $\dfrac{2}{3}\ln(3x+2) + c$

(e) $\dfrac{1}{10}\ln(5x^2 - 2) + c$ (f) $\dfrac{1}{2}e^{2x+1} + c$

(g) $\dfrac{1}{16}(4x+1)^4 + c$ (h) $\dfrac{-1}{(1+x)} + c$

(i) $\dfrac{1}{5}\ln(5x-3) + c$ (j) $\dfrac{1}{2}e^{x^2} + c$

(k) $\dfrac{1}{1.3}(1+x)^{1.3} + c$ (l) $-\ln(\cos x) + c$

(m) $\dfrac{2}{3}(x^2+x+3)^{1.5} + c$ (n) $\ln(x^2 + 3x - 5) + c$

(o) $\dfrac{1}{24}(4t-11)^6 + c$ (p) $\dfrac{1}{2(3-2r)} + c$

(q) $\dfrac{1}{2}\sin(2y+\pi) + c$ (r) $\dfrac{1}{2}\ln(1+t^2) + c$

(s) $-\dfrac{1}{3}\cos^3(\theta) + c$ (t) $\dfrac{1}{10}\sin^5(2\theta) + c$

2. (a) $-\dfrac{1}{3}$ (-0.33) (b) $\dfrac{1}{5}\ln(6)$ (0.3584)

(c) $\dfrac{1}{2}\ln\left(\dfrac{5}{3}\right)$ (0.2554) (d) $\left(5\sqrt{5}-1\right)/3$ (3.3934)

(e) $\dfrac{2}{3}\ln(3)$ (0.7324) (f) $\dfrac{1}{2}(e^1-1)$ (0.8591)

(g) $\dfrac{1}{3}(e^1-e^{-8})$ (0.9060) (h) 34

Exercise 7G

(a) $\dfrac{1}{2}\arcsin(2x) + c$ (b) $\dfrac{1}{2}\arctan(2t) + c$

(c) $\arcsin\left(\dfrac{x}{3}\right) + c$ (d) $\dfrac{1}{3}\arctan(3x) + c$

(e) $\dfrac{\pi}{6}(0.5236)$ (f) $\dfrac{\pi}{4}(0.7854)$

Exercise 7H

1. (a) $-x\cos x + \sin x + c$ (b) $xe^x - e^x + c$

(c) $\dfrac{1}{2}x^2 e^{2x} - \dfrac{1}{2}xe^{2x} + \dfrac{1}{4}e^{2x} + c$ (d) $-\dfrac{1}{3}x\cos 3x + \dfrac{1}{9}\sin 3x + c$

(e) $t^2 e^t - 2te^t + 2e^t + c$ (f) $\dfrac{1}{2}x^2\ln x - \dfrac{1}{4}x^2 + c$

(g) $x\ln(x) - x + c$

2. (a) 1 (b) -0.5 (c) 8.6328 (d) 0.1905

(e) $(1+e^{\pi/2})/2$ (2.9052) (f) $\dfrac{3}{13}(e^{-2\pi/3}+1)$ (0.2592)

354   **Answers**

Exercise 7I

(a) $\ln(x-1) - \ln(x+1) + c$

(b) $\dfrac{1}{5}\ln(2x-1) - \dfrac{1}{5}\ln(x+2) + c$

(c) $\dfrac{5}{4}\ln(x-3) - \dfrac{5}{4}\ln(x+1) + c$

(d) $3\ln(t-3) - 3\ln(t-2) + c$

(e) $\dfrac{1}{4}\ln(2v-1) + \dfrac{1}{4}\ln(2v+3) + c$

(f) $5\ln(5) + 3\ln(3) - 16\ln(2)$ (0.2527)

(g) $\dfrac{23}{6}\ln(x+3) - \dfrac{3}{2}\ln(x-1) - \dfrac{1}{3}\ln(x) + c$

(h) $\dfrac{5}{4}\ln(2) - \dfrac{3\pi}{8}(-0.3117)$

(i) $\dfrac{2}{\sqrt{19}}\arctan\left((2x-9)/\sqrt{19}\right)$

(j) $\ln(3) - 2\ln(2) + 0.5$ (0.2123)

Exercise 7J

1. (a) $\dfrac{1}{2}\sin^2 x + c$; or $-\dfrac{1}{2}\cos^2 x + c$; or $-\dfrac{1}{4}\cos 2x + c$

(b) $\ln(x^3-3) + c$

(c) $\dfrac{1}{2}e^{2x} + \dfrac{1}{4}\cos 4x + c$

(d) 2

(e) $-\dfrac{1}{4}x\sin(2x) + \dfrac{1}{4}\cos 2x + \dfrac{1}{4}x^2 + c$

(f) $\dfrac{1}{2}\arctan\left(\dfrac{1}{2}\right)$ (0.2318)

(g) $\dfrac{1}{4}\ln(3)$ (0.2747)

(h) $\dfrac{1}{2}\ln(4+i^2) + c$

(i) $\dfrac{1}{8}(e^4-1)$ (6.70)

(j) $\ln(t-2) - \ln(t+3) + c$

(k) $\dfrac{1}{2}u^2 + 2\ln(u) + c$

(l) $\dfrac{1}{2}\arcsin(v) + \dfrac{1}{2}v\sqrt{1-v^2} + c$

(m) $\dfrac{1}{5}u^5 + \dfrac{1}{3}u^3 + c$

(n) $\dfrac{1}{4}(e^7-e^{-1})$ (274.07)

(o) $(2-u^2)\cos(u) + 2u\sin(u) + c$

(p) $\dfrac{1}{3}te^{3t} - \dfrac{1}{9}e^{3t} + c$

2. 875 metres

3. $T = \dfrac{k}{2(2t+3)} + T_0$

where k is the constant of proportionality and T_0 is a constant of integration

4. $V = \dfrac{15}{2}x^2 - 20x + c$

5. $y = \ln(x - 2) - \ln(x - 1) + 1 + \ln(2)$

6. total area $= 16.5$

7. area $= \dfrac{1}{6}$ (0.167)

8. 26.3856 (using right ends of intervals)

9. (a) area $= 0.5$
    (b) $\displaystyle\int_1^2 (2x - 3)dx = 0$

10. $t = 0.5919\dfrac{u}{g}$

11. $H = 0.4472$

Exercise 8A

1. (a) 1.89549, 0, −1.89549   (b) −2.1663
    (c) 1.85722, 4.5364   (d) 0.7391
    (e) −1, 1.3532   (f) 0.5671

2. (a) 2.1544   (b) 1.5157   (c) −1.5850, 1.5850

3. The method does not converge for $x = -1$ and $x = 1$

4. (0.1444, 1.0719) Maximum
    (3.2617, −16.4608) Minimum

5. $K = 0.1594$

Exercise 8B

1. (a) $1 + x + \dfrac{x^2}{2}$ (b) $1 - \dfrac{x^2}{2} + \dfrac{x^4}{24}$ (c) $x^2 - \dfrac{x^6}{6} + \dfrac{x^{10}}{120}$

    (d) $1 - x + x^2$ (e) $x + \dfrac{x^3}{3} + \dfrac{2x^5}{15}$ (f) $1 - \dfrac{x}{2} - \dfrac{x^2}{8}$

    (g) $-\dfrac{x^2}{2} - \dfrac{x^4}{12} - \dfrac{x^6}{45}$ (h) $1 + x^2 + \dfrac{x^4}{2}$

Exercise 8C

1. (a) $0.5 - 0.866\left(x - \dfrac{\pi}{3}\right) - 0.25\left(x - \dfrac{\pi}{3}\right)^2$

    (b) $1 + 2\left(x - \dfrac{\pi}{4}\right) + 2\left(x - \dfrac{\pi}{4}\right)^2$ (c) $1 - (x-1) + (x-1)^2$

2. (a) $2x - \dfrac{4}{3}x^3 + \dfrac{4}{15}x^5$ (b) $3x - \dfrac{9}{2}x^3 + \dfrac{81}{40}x^5$

3. (b) $-\dfrac{x^2}{2} - \dfrac{x^4}{12} - \dfrac{x^6}{45} - \dfrac{17x^8}{2520} - \dfrac{31x^{10}}{14175}$ (c) -0.05003

4. 1.4408

5. 1.4618

Exercise 8D

1. (a) trapezoidal: 2.9814
         Simpson: 2.9274

    (b) trapezoidal: 1.5656
         Simpson: 1.5708

    (c) trapezoidal: 0.4058
         Simpson: 0.4055

2. (a) 0.746546   (b) 0.746528   (c) 0.746526
    The answer is 0.7465 to four decimal places

3. 0.845 (using 32 subintervals)

4. trapezoidal: 8.81
    Simpson: 8.83
    Percentage error: 5.8%

Exercise 9A

(a) $\begin{bmatrix} 1 & -1 \\ 2 & 3 \end{bmatrix}\begin{bmatrix} x \\ y \end{bmatrix} = \begin{bmatrix} 4 \\ 1 \end{bmatrix}$

(b) $\begin{bmatrix} -2 & 1 \\ 1 & -1 \end{bmatrix}\begin{bmatrix} x \\ y \end{bmatrix} = \begin{bmatrix} -6 \\ 4 \end{bmatrix}$

(c) $\begin{bmatrix} 3 & -6 & 1 \\ -2 & 1 & -3 \\ 1 & 1 & 1 \end{bmatrix}\begin{bmatrix} x \\ y \\ z \end{bmatrix} = \begin{bmatrix} 7 \\ 2 \\ 0 \end{bmatrix}$

(d) $\begin{bmatrix} 4 & -1 & 1 \\ 3 & 1 & -2 \\ -1 & -4 & 1 \end{bmatrix}\begin{bmatrix} x_1 \\ x_2 \\ x_3 \end{bmatrix} = \begin{bmatrix} 1 \\ -3 \\ 5 \end{bmatrix}$

(e) $\begin{bmatrix} 1 & -3 & 1 & 4 & -1 \\ -1 & 0 & -4.1 & 0 & 2 \\ 0.3 & -0.7 & 0 & 4.1 & -1 \\ 1.4 & -1 & 0 & 3.1 & 0 \\ 3 & 1 & -1 & 2 & -1 \end{bmatrix}\begin{bmatrix} x_1 \\ x_2 \\ x_3 \\ x_4 \\ x_5 \end{bmatrix} = \begin{bmatrix} 0 \\ 1.3 \\ 2.7 \\ 0.4 \\ -3.5 \end{bmatrix}$

Exercise 9B

1. (a) $\begin{bmatrix} 1 & 5 & 5 \\ -1 & -2 & 3 \\ -1 & 2 & 2 \end{bmatrix}$ (b) Impossible (c) $\begin{bmatrix} 0 & -3 & -9 \\ 3 & 0 & -6 \\ 0 & -3 & -6 \end{bmatrix}$

(d) $\begin{bmatrix} 1 & 1 \\ 1 & 1 \end{bmatrix}$ (e) $\begin{bmatrix} -1 & 1 \\ 1 & -1 \end{bmatrix}$ (f) $\begin{bmatrix} 2 & 5 & -5 \\ 3 & -4 & -4 \\ -2 & -1 & -6 \end{bmatrix}$

(g) Impossible (h) $\begin{bmatrix} -1 & 2 & 1 \\ 1 & 3 & 0 \end{bmatrix}$ (i) Impossible

(j) Impossible (k) $\begin{bmatrix} -2 & -7 \\ -2 & -7 \\ -1 & -4 \end{bmatrix}$ (l) $\begin{bmatrix} 11 & 2 & 0 & 6 \\ 8 & -1 & -1 & 1 \\ 7 & 2 & 1 & 4 \end{bmatrix}$

(m) Impossible (n) $\begin{bmatrix} 1 & -2 & 5 \\ -2 & -7 & 0 \end{bmatrix}$ (o) $\begin{bmatrix} 0 & 1 \\ 1 & 0 \end{bmatrix}$

(p) Impossible (q) $\begin{bmatrix} 1 & 3 & 0 \\ -1 & 2 & 1 \end{bmatrix}$ (r) $\begin{bmatrix} 0 & 1 \\ 1 & 2 \\ -1 & -3 \end{bmatrix}$

2. $Bx = \begin{bmatrix} x_1 + 3x_2 \\ -x_1 + 2x_2 + x_3 \end{bmatrix}$ $Cx = \begin{bmatrix} x_1 + 4x_2 + 2x_3 \\ -2x_2 + x_3 \\ -x_1 + x_2 \end{bmatrix}$ $Ex = \begin{bmatrix} x_2 + 3x_3 \\ -x_1 + 2x_3 \\ x_2 + 2x_3 \end{bmatrix}$

3.   (a) $\begin{bmatrix} -1 & -3 & 1 \\ -1 & 2 & 1 \\ 1 & 0 & 2 \end{bmatrix}$ (b) $\begin{bmatrix} -3 & -7 & 9 \\ -5 & 6 & 7 \\ 3 & 2 & 10 \end{bmatrix}$ (c) $\begin{bmatrix} \frac{1}{2} & 3 & 4 \\ -1 & -1 & \frac{5}{2} \\ -\frac{1}{2} & \frac{3}{2} & 2 \end{bmatrix}$

4. $a = 0, b = 0$

Exercise 9C

1.   (a) $\begin{bmatrix} \frac{2}{17} & \frac{5}{17} \\ \frac{3}{17} & -\frac{1}{17} \end{bmatrix}$ (b) $\begin{bmatrix} 2 & -1 \\ \frac{1}{2} & 0 \end{bmatrix}$ (c)   No inverse

      (d) $\begin{bmatrix} \frac{70}{61} & \frac{40}{61} \\ -\frac{100}{61} & \frac{30}{61} \end{bmatrix}$

2.   (a) $x = \dfrac{9}{17}, \quad y = \dfrac{5}{17}$ (b) $x = \dfrac{116}{61}, \quad y = -\dfrac{96}{61}$

      (c) $x = 2, y = 3$ (d) $x = \dfrac{dr - bs}{ad - bc}, \quad y = \dfrac{as - cr}{ad - bc}$

3. $x = \dfrac{19}{18}, \quad y = -\dfrac{5}{6}, \quad z = \dfrac{10}{9}$

4. $x_1 = -0.1, \ x_2 = -1.9, \ x_3 = 8, \ x_4 = 1.9$

5.   (B) (i) $\dfrac{1}{9}\begin{pmatrix} 4 & -1 \\ 1 & 2 \end{pmatrix}$ (ii) $\dfrac{1}{14}\begin{pmatrix} 3 & -1 & 5 \\ 5 & 3 & -1 \\ -1 & 5 & 3 \end{pmatrix}$

      (iii) $\dfrac{1}{9}\begin{pmatrix} -3 & 12 & 0 & -21 \\ 1 & -5 & 2 & 10 \\ 1 & 1 & -1 & 1 \\ 3 & -3 & 0 & 3 \end{pmatrix}$ (iv)   singular matrix, no inverse

      (C) (i)   9   (ii)   14   (iii)   −27   (iv)   0

(D) $x_1 = -1, x_2 = 1, x_3 = 2$

(E)   (i) $x_1 = 2, x_2 = -1, x_3 = -1$
      (ii) $x = 1, y = 0, z = 1, w = 0$
      (iii) $x_1 = 1, x_2 = 1, x_3 = 1, x_4 = -1$
      (iv) $x_1 = -0.1, x_2 = -1.9, x_3 = 8, x_4 = 1.9$

Exercise 10A

1.

Complex Number	Real Part	Imaginary Part	Complex Conjugate
$2 - 3i$	2	-3	$2 + 3i$
$6 + 2i$	6	2	$6 - 2i$
$1.73 - 2.19i$	1.73	-2.19	$1.73 + 2.19i$
$1.7 + 4.6i$	1.7	4.6	$1.7 - 4.6i$
$-5.17 + 1.03i$	-5.17	1.03	$-5.17 - 1.03i$
$-4i$	0	-4	$4i$
$17i$	0	17	$-17i$
$x - yi$	x	$-y$	$x + yi$
$-p + qi$	$-p$	q	$-p - qi$

2. $x = -1+\sqrt{2}i, x = -1-\sqrt{2}i$

$x = 3i, x = -3i$

$x = \dfrac{1}{4}+\dfrac{\sqrt{7}}{4}i, x = \dfrac{1}{4}-\dfrac{\sqrt{7}}{4}i$

$x = -\dfrac{1}{3}+\dfrac{\sqrt{5}}{3}i, x = -\dfrac{1}{3}-\dfrac{\sqrt{5}}{3}i$

4.   (a) $x^2 - 4x + 13 = 0$
     (b) $100x^2 - 40x + 29 = 0$
     (c) $x^2 - 6x + 34 = 0$
     (d) $x^2 - 2ax + (a^2+b^2) = 0$

Exercise 10B

1.   (a)   modulus $= 1$, argument $= \dfrac{\pi}{2}$

     polar form $= \cos\dfrac{\pi}{2} + i\sin\dfrac{\pi}{2}$

(b)   modulus = $\sqrt{2}$, argument = $\dfrac{\pi}{4}$

  polar form = $\sqrt{2}\left(\cos\dfrac{\pi}{4}+i\sin\dfrac{\pi}{4}\right)$

(c)   modulus = $\sqrt{13}$, argument = -0.983
  polar form = $\sqrt{13}(\cos(-0.983)+i\sin(-0.983)$
   $= \sqrt{13}(\cos 0.983 - i\sin 0.983)$

(d)   modulus = $\sqrt{17}$, argument = -1.816
  polar form = $\sqrt{17}(\cos(-1.816)+i\sin(-1.816))$
   $= \sqrt{17}(\cos 1.816 - i\sin 1.816)$

(e)   modulus = 5, argument = -2.214
  polar form = $5(\cos(-2.214) + i\sin(-2.214))$
   $= 5(\cos 2.214 - i\sin(2.214))$

(f)   modulus = 13, argument = 1.966
  polar form = $13(\cos 1.966 + i\sin 1.966)$

(g)   modulus = 1, argument = 0
  polar form = $1(\cos 0 + i\sin 0)$

(h)   modulus = 3, argument = $-\dfrac{\pi}{2}$

  polar form = $3\left(\cos\left(-\dfrac{\pi}{2}\right)+i\sin\left(-\dfrac{\pi}{2}\right)\right)$

   $= 3\left(\cos\dfrac{\pi}{2} - i\sin\dfrac{\pi}{2}\right)$

(i)   modulus = 4, argument = π
  polar form = $4(\cos(-\pi) + i\sin(-\pi))$
   $= 4(\cos\pi - i\sin\pi)$

(j)   modulus = 2, argument = -1.047
  polar form = $2(\cos(-1.047) + i\sin(-1.047))$
   $= 2(\cos 1.047 - i\sin 1.047)$

(k)   modulus = 2, argument = –2.094
      polar form = 2(cos(–2.094) + isin(–2.094))
                 = 2(cos2.094 – isin2.094)

(l)   modulus = 6.35, argument = –1.36
      polar form = 6.35(cos(–1.36) + isin(–1.36))
                 = 6.35(cos1.36 – isin1.36)

2.   (a) $x = -1 + i, x = -1 - i$

$$x = \sqrt{2}\left(\cos\frac{3\pi}{4} + i\sin\frac{3\pi}{4}\right), x = \sqrt{2}\left(\cos\frac{3\pi}{4} - i\sin\frac{3\pi}{4}\right)$$

(b) $x = \sqrt{8}i, x = -\sqrt{8}i$

$$x = \sqrt{8}\left(\cos\frac{\pi}{2} + i\sin\frac{\pi}{2}\right), x = \sqrt{8}\left(\cos\frac{\pi}{2} - i\sin\frac{\pi}{2}\right)$$

3.   (a) 5   (b) –0.927   (c) 5.099
     (d) 9.19   (e) –1.96   (f) 111.803

Exercise 10C

1.   (a) $3 + 5i, -3 - 3i, -4 + 3i, \dfrac{4}{25} + \dfrac{3}{25}i, -1, -6 - 5i$

(b) $-2 + 3i, -4 + 5i, 1 + 7i, -\dfrac{7}{2} + \dfrac{1}{2}i, -7 - 24i, -11 + 14i$

(c) $2, 2i, 2, i, 2i, 1 + 5i$

(d) $-5 - 3i, -7 + i, -8 + 11i, -\dfrac{4}{5} - \dfrac{13}{5}i, 35 + 12i, -20 + i$

(e) $8 + i, -2 + 3i, 17 + 7i, \dfrac{1}{2} + \dfrac{1}{2}i, 5 + 12i, -1 + 8i$

(f) $2 + 8i, -8 - 16i, 33 - 56i, -\dfrac{63}{169} + \dfrac{16}{169}i, 7 + 24i, -19 - 36i$

2.   (a) $2 - 3i$ (b) $3 + 2i$ (c) $\dfrac{3}{13} + \dfrac{2i}{13}$ (d) $\dfrac{3}{13} - \dfrac{2}{13}i$

4.   (a) $a = 3\cos x, b = -3\sin x$

(b) $a = \dfrac{2\cos x + \sin x + 2}{2(\cos x + 1)}, b = \dfrac{-(\cos x - 2\sin x + 1)}{2(\cos x + 1)}$

5. $\text{Re}(z) = \dfrac{1}{5}, \text{Im}(z) = \dfrac{7}{5}$

6. (b) $|z^n| = |z|^n$, $\arg(z^n) = n\arg(z)$

    $z^n = r^n(\cos(n\theta) + i\sin(n\theta))$

Exercise 10D

1. (a) $5(\cos 0.927 + i\sin 0.927) = 5e^{0.927i}$

    (b) $\sqrt{2}\left(\cos\left(\dfrac{\pi}{4}\right) - i\sin\left(\dfrac{\pi}{4}\right)\right) = \sqrt{2}e^{-\frac{\pi}{4}i}$

    (c) $\cos\left(\dfrac{5\pi}{6}\right) - i\sin\left(\dfrac{5\pi}{6}\right) = e^{-\frac{5\pi}{6}i}$

    (d) $13(\cos 1.176 - i\sin 1.176) = 13e^{-1.176i}$

    (e) $2\sqrt{2}\left(\cos\dfrac{3\pi}{4} + i\sin\dfrac{3\pi}{4}\right) = 2\sqrt{2}e^{\frac{3\pi}{4}i}$

    (f) $0.762(\cos 1.976 - i\sin 1.976) = 0.762e^{-1.976i}$

    (g) $\dfrac{1}{5}(\cos 0.93 - i\sin 0.93), \dfrac{1}{5}e^{-0.93i}$

    (h) $\dfrac{1}{\sqrt{5}}(\cos 1.11 - i\sin 1.11), \dfrac{1}{\sqrt{5}}e^{-1.11i}$

    (i) $0.109(\cos 0.391 + i\sin 0.391), 0.109e^{0.391i}$

2. (a) -1 (b) $3i$ (c) -1 (d) $1.53 + 1.29i$
    (e) $3.980 - 0.399i$ (f) $0.636 + 0.63i$

3. (a) modulus = 4, argument = $\dfrac{\pi}{2}$ (b) modulus = 27, argument = $\dfrac{\pi}{2}$

    (c) modulus = 108, argument = π (d) modulus = $\dfrac{4}{27}$, argument = 0

4. (a) modulus = 16, argument = 1.4   (b) modulus = 0.125, argument = -0.3

    (c) modulus = 2, argument = 1.1   (d) modulus = 128, argument = 1.7

5.   5(cos0.927 − $i$sin0.927), 1526(cos0.719 − $i$sin0.719)

6. $\sqrt{2}\left(\cos\dfrac{\pi}{4}+i\sin\dfrac{\pi}{4}\right)$,   64(cos3π + $i$sin3π) = −64

7.   cos3θ = cos^{3}θ − 3sin^{2}θ cosθ = 4cos^{3}θ − 3cosθ
      sin3θ = 3sinθ cos^{2}θ − sin^{3}θ = 3sinθ − 4sin^{3}θ

Exercise 10E

		Square Roots	Cube Roots
1.	(a)	1.09868 − 0.455089i −1.09868 + 0.455089i	1.08421 − 0.290514i −0.290514 + 1.08421i −0.793700 − 0.793700i
	(b)	1.45534 + 0.343560i −1.45534 − 0.343560i	1.29207 + 0.201294i −0.820363 + 1.01832i −0.471711 − 1.21961i
	(c)	1.22474 + 1.22474i −1.22474 − 1.22474i	1.24902 + 0.721124i −1.24902 + 0.721124i 0 − 1.44224i
	(d)	2 + 0i −2 + 0i	1.58740 −0.793700 + 1.37472i −0.793700 − 1.37472i
	(e)	0.707106 + 1.22474i −0.707106 − 1.22474i	0.965155 + 1.09112i −1.18393 − 1.09112i 0.218782 + 1.09112i
	(f)	2 − 3i −2 + 3i	1.86444 − 1.43269i 0.308530 + 2.33100i −2.17297 − 0.898307i

2.   (a) $2^{\frac{1}{5}}\left(\cos\left(\dfrac{-\pi/6+2k\pi}{5}\right)+i\sin\left(\dfrac{-\pi/6+2k\pi}{6}\right)\right)$ $k = 0,1,2,3,4$

      (b) $2^{\frac{2}{3}}\left(\cos\left(\dfrac{\pi/4+2k\pi}{3}\right)+i\sin\left(\dfrac{\pi/4+2k\pi}{3}\right)\right)^{4}$

            $= 2^{\frac{2}{3}}\left(\cos\left(\dfrac{\pi+8k\pi}{3}\right)+i\sin\left(\dfrac{\pi+8k\pi}{3}\right)\right)$ $k = 0,1,2$

(c) $5^{-\frac{2}{3}}\left(\cos\left(\dfrac{2.214+2k\pi}{3}\right)+i\sin\left(\dfrac{2.214+2k\pi}{3}\right)\right)^{-2}$

$\qquad = 5^{-\frac{2}{3}}\left(\cos\left(\dfrac{4.428+4k\pi}{3}\right)-i\sin\left(\dfrac{4.428+4k\pi}{3}\right)\right)\quad k=0,1,2$

(d) $\cos\left(\dfrac{(2k+1)\pi}{6}\right)+i\sin\left(\dfrac{(2k+1)}{6}\right)\quad k=0,1,2,3,4,5$

3. $-0.471711 + 1.21961i$
    $-0.820363 - 1.01832i$

4. (a) $1.21589 + 0.503639i$
        $-0.503639 + 1.21589i$
        $-1.21589 - 0.503639i$
        $0.503639 - 1.21589i$

    (b) $1.11819 - 0.097829i$
        $0.643817 + 0.919467i$
        $-0.474372 + 1.01729i$
        $-1.11819 + 0.97829i$
        $-0.643817 - 0.919467i$
        $0.474372 - 1.01729i$

Index

Other books in the "Learning with Computer Algebra" Series
Published by Chartwell-Bratt. Series editor John Berry, University of Plymouth

Learning Mathematics through DERIVE
J. S. Berry, E. Graham, A. J. P. Watkins
The DERIVE version of "Learning Mathematics through the TI-92."
  ISBN 0-86238-461-3, 370 pages, 1996

Learning Numerical Analysis through DERIVE
Terence Etchells, John Berry
This book covers the major numerical methods, and their analysis, for first courses at college and undergraduate level. The relative merits of each method are covered both analytically, providing a thorough grounding in the algebraic approach, and practically, through the tried and tested computer lab-based activities.
DERIVE provides a platform on which to quickly and accurately perform many complicated numerical calculations. Also, DERIVE's ability to algebraically manipulate expressions and perform calculus operations, enhances the investigation of the convergence of numerical methods.
Each chapter includes the development and algebraic analysis of the methods, lab-based activities, ideas for coursework, case studies, exercises and solutions. Free supporting utility files are downloadable via Chartwell-Bratt's web server.
Chapter 1 introduces the basic tool of numerical methods, which is recurrence relations, their solution and ill-conditioning problems. In chapter 2 we use recurrence relations methods that are used in solving equations. Chapter 3 deals with the approximation of functions by polynomials, and in particular the Taylor Polynomial, which is then used extensively in chapter 4 to analyse the errors associated with numerical methods.
Chapters 5 and 6 deal with numerical approaches to the calculus of differentiation and integration. In chapter 7 we introduce and analyse numerical methods of solving differential equations.   ISBN 0-86238-468-0, 239 Pages, 1997

Learning Linear Algebra through DERIVE
B. Denton
Using DERIVE, this book reinforces theoretical knowledge while making applications more realistic. It can also be used with packages such as Matlab, Maple, Mathematica and Macsyma, requiring only a few adjustments. The book covers a two-semester course, and will be essential reading for undergraduate students of linear algebra.
CONTENTS: Introduction to matrices; Vectors with applications to geometry; Systems of linear equations; Vector spaces; Linear transformation, Eigenvectors and eigenvalues, Conclusions; Solutions.
ISBN 0-86238-466-4, 296 pages, 1995

Learning Modelling with DERIVE

S. Townend, D. Pountney

Develops undergraduate mathematical modelling skills with the support of DERIVE. The book provides students with the opportunity to develop both their problem-solving and IT skills through the fully integrated use of DERIVE. The authors' experience of teaching modelling at Liverpool John Moores University shows that the use of DERIVE enhances the students' modelling expertise.

- Gently guides the reader through the problem formulating, solution and revision stages of modelling.

- Provides a wide range assortment of case studies, some of which are developed fully while others have only hints provided. A wide-ranging set of modelling problems is also provided.

CONTENTS: Introduction; Geometric and trigonometric models; Algebraic models; Optimisation-based models; Statistical and simulations models; The techniques of dimensional analysis.

ISBN 0-86238-467-2, 256 pages, 1995

Other books on using DERIVE in education

Mathematical Activities with DERIVE

E. Graham, J. S. Berry, A. J. P. Watkins (eds)

These mathematical activities using DERIVE, were selected to illustrate the variety of ways in which DERIVE can be used to enhance the teaching and learning of mathematics. The activities contain many ideas that could easily be applied in different areas of mathematics, and which will act as a stimulus for developing your own tasks.

They are pitched at a variety of levels, from secondary school, through college to university. Several activities contain ideas that can be used at different levels. For example, the fractals section contains work on transformations that could be used with some school students, while some of the later activities are only suitable for undergraduates.

Each chapter includes a Rationale, a Description of the Activities, Comments on the Use of the Activities, to supplement the activities themselves. This allows readers to identify the key aspects of each activity and the advantages associated with its use, as well as any problems that might be experienced.

Free utility files to support the activities are available from the authors.

ISBN 0-86238-478-8, 216 pages, 1997, A4 spiral bound

Improving Mathematics Teaching with DERIVE: a guide for teachers

B. Kutzler

Computer algebra systems are a new generation of mathematical tools. They let you process symbolic data (e.g. formulae, expressions, equations) as fast and accurately as a scientific calculator processing numbers. This makes computer algebra systems the biggest challenge to mathematics education in history.

This book is an introduction to using the computer algebra system DERIVE for teaching mathematics. The author explains the "scaffolding didactics" as a method of using computer algebra systems for teaching mathematics alongside existing, traditional, curricula. He also develops a contemporary, information technology age, concept for teaching mathematics by discussing the following questions:

  * Which established topics can be omitted?
  * Which traditional topics require more attention?
  * Which new topics may be added?
  * Which new classroom techniques can be applied?
  * Which new teaching goals should be considered?

The book also gives ready-made recipes for using the system in teaching certain topics, gives tips and tricks for using DERIVE, describes a useful non-documented product feature, and provides a look "inside" computer algebra systems. The book is written for mathematics teachers at schools, colleges and universities.

Bernhard Kutzler holds a PhD in mathematics, has written numerous publications in the field of computer algebra, and teaches at the University of Linz. He is active on a national and an international level and has been involved in various projects about using DERIVE for teaching mathematics.

CONTENTS: Introduction; What DERIVE can do; DERIVE in Traditional Mathematics Teaching; The Future of Mathematics Teaching; Examples for Teaching; Tips and Tricks; Your own system - user defined menus; How Does a Computer Algebra System Work? DERIVE and 250 of history - Quo Vadis? Index.

ISBN 0-86238-422-2, 185 pages, 1996

Elementary Linear Algebra with DERIVE: an integrated text
R. J. Hill, T. A. Keagy

Linear algebra is an area of mathematics which blends an elegant theoretical structure with a rich supply of applications, but the study of linear algebra has historically been burdened with tedious calculations which have distracted the learner and discouraged the user.

In this text extensive use is made of Derive software to simplify calculations and focus on the beauty of the structure of linear algebra and its associated applications. The elementary properties of matrices, vector spaces, linear transformations, and determinants are presented in a style that promotes an understanding and appreciation of the order and logic of the theory (without the unnecessary frustration of long, time-consuming calculations).

Since this early material is covered more efficiently through Derive, more emphasis can be spent on the study of eigenvalues, eigenvectors, matrix analysis, and applications in diverse areas such as least square approximations, Markof processes, difference equations, and linear systems

of differential equations with constant coefficients. To support the learner, the text includes 180 example problems worked in detail, almost 600 Derive statements illustrating the use of the software, and more than 600 exercises with solutions.
ISBN 0-86238-403-6, 392 pages, 1995.

Other mathematics technology books (send for full catalogue)

Computer Algebra in Mathematics Education: State of the Art
J. Berry, M. Kronfellner, B. Kutzler, J. Monaghan
The calculator changed the teaching of numerical mathematics forever. Now, low-cost computer algebra software, and the Texas Instruments TI-92 calculator, are revolutionising teaching of advanced mathematics. As with any powerful tool, guidance in its use is necessary. This book is a collaboration between many of the world's leading experts on using computer algebra systems to teach mathematics. It distills their combined expertise into one volume and represents the state of the art.   ISBN 0-86238-430-3, 168 pages, 1997.

Mathematical Activities with Computer Algebra: a photocopiable resource book
T. A. Etchells, M. Hunter, J. Monaghan, S. Pozzi, A. Rothery
Computer algebra systems will have a significant impact on the way people carry out their mathematics. This photocopiable resource book is the first of a new generation of support materials for the educational use of computer algebra. Designed to be used with **any computer algebra system**, the authors go beyond mere button pressing and show how to harness the power of computer algebra systems for educational purposes.

The book is aimed at students in the over-16 age range in sixth forms, colleges and first year higher education in universities, in the UK and USA. The choice of topics is compatible with A level mathematics in the UK, SCAA guidelines and NCTM standards in the USA.

Concepts are illustrated, techniques and methods presented, and modelling and applications are explained. Activity Worksheets, Help Sheets and Teaching Notes cover a wide range of mathematical topics, including: functions and graphs, differentiation, integration, sequences and series, vectors and matrices, mechanics, trigonometry, and numerical methods. The activities illustrate the potential impact of this revolutionary technology and will help teachers and lecturers make a start in implementing its use.

The design of the book enables it to be used flexibly for class, lab, group or individual work. Appendices give overviews of DERIVE, Maple, Mathematica, Theorist (MathPlus), Macsyma and the TI-92 calculator, showing how each implements the key commands.

What the book offers:-
* 23 photocopiable student worksheets
* 23 photocopiable optional student help sheets
* detailed teaching notes and solutions

* an approach which permits use of any computer algebra system
* a range of mathematical topics in the 16+ curriculum
* activities which easily align with standard teaching topics

Organisation and content:-

The activities are at different levels of difficulty, making it easier to identify activities for the particular needs of any group of students. Some activities cover mathematical techniques and concepts and many relate to an applicable mathematics or modelling context. In some cases, suggested extensions to activities provide ample material for an extended project.
ISBN 0-86238-405-2, 126 pages, 1997.

Technology in Mathematics Teaching: a bridge between teaching and learning
L. Burton, B. Jaworski
The book is divided into six sections: Setting the Scene; Why Technology; Using Calculators; How may technology support school algebra?; Advanced Mathematics - Some Perspectives; Innovative Uses of Technology.
ISBN 0-86238-401-X, 496 pages, 1995.

Numerical Analysis and Spreadsheets
D. McLaren
Using spreadsheets as a powerful tool to discover numerical analysis at college and undergraduate level. The text includes examples to illustrate the various methods discussed and includes appropriate sample spreadhseets. Many are structured so that the reader has the exercise of finishing them in order to see the results mentioned in the text. It is intended that readers' curiosity will lead them to explore beyond the given problems.

The book is based on a second year course taught by the author in the La Trobe University School of Mathematics. The mathematical knowledge assumed is that gained from first-year mathematics, namely basic calculus and elementary matrix algebra.

CONTENTS: Introduction to the spreadsheet; Convergence of sequences; Simultaneous linear equations; Solution of non-linear equations; Numerical integration; Ordinary differential equations; Partial differential equations.
ISBN 0-86238-431-1, 250 pages, 1997.